AF518298

Diseases of Vegetable, Spices and Plantation Crops

The Author

Dr R S Mishra, Associate Professor (Senior Scale), Plant Pathology at N.D. University of Agriculture and Technology, Kumarganj, Faizabad, U.P. He has more than 17 years experience in teaching and research. He obtained M.Sc. (Ag.) degree from Allahabad Agricultural Institute Allahabad and Ph.D. degree from Jamia Hamdard University, Delhi. Dr. Mishra had worked on Interactive effect of elevated CO_2 and moisture stress on crop plants at IARI, New Delhi. He has been involved in research on Agro-meteorology and Agro-Advisory Services. He obtained training from CRIDA, Hyderabad, IITM, Pune and IARI, New Delhi on the different aspect of Agro meteorology and plant disease management. Dr. Mishra developed protocol for Mushroom cultivation and management practices of *Coprinus* spp. in paddy stray mushroom. Presently, Dr. Mishra is actively engaged in teaching and research of Vegetable and Spices Diseases. Dr. Mishra has invited membership of New York Academy of Sciences. He has to his credit more than 80 publications including research papers, books, book chapter, manuals, and popular articles. Dr. Mishra has developed four varieties of Haldi. He is the recipient of several Awards of professional Scientific Societies.

Diseases of Vegetable, Spices and Plantation Crops

– Author –

R S Mishra

2019

Daya Publishing House®

A Division of

Astral International Pvt. Ltd.

New Delhi – 110 002

© 2019 AUTHOR

ISBN: 978-93-88173-16-2 (HB)

Publisher's Note:

Every possible effort has been made to ensure that the information contained in this book is accurate at the time of going to press, and the publisher and author cannot accept responsibility for any errors or omissions, however caused. No responsibility for loss or damage occasioned to any person acting, or refraining from action, as a result of the material in this publication can be accepted by the editor, the publisher or the author. The Publisher is not associated with any product or vendor mentioned in the book. The contents of this work are intended to further general scientific research, understanding and discussion only. Readers should consult with a specialist where appropriate.

Every effort has been made to trace the owners of copyright material used in this book, if any. The author and the publisher will be grateful for any omission brought to their notice for acknowledgement in the future editions of the book.

All Rights reserved under International Copyright Conventions. No part of this publication may be reproduced, stored in a retrieval system, or transmitted in any form or by any means, electronic, mechanical, photocopying, recording or otherwise without the prior written consent of the publisher and the copyright owner.

Published by : **Daya Publishing House®**
A Division of
Astral International Pvt. Ltd.
– ISO 9001:2015 Certified Company –
4736/23, Ansari Road, Darya Ganj
New Delhi-110 002
Ph. 011-43549197, 23278134
E-mail: info@astralint.com
Website: www.astralint.com

Digitally Printed at : **Replika Press Pvt. Ltd.**

This copy is for sale in Indian Sub-Continent.

Foreword

Vegetable and spices is an important crop of India and is grown throughout the country under varied Agro-climatic condition. India has the distinction of being the top producer and exporter of vegetable and spices in the world market. In spite of such an impressive figure, the country has experienced stagnant decline in per capita availability of these horticultural crops. The decline is due to several regions but susceptibility to many diseases and insect pests are important factors. The present productivity is 168592.5 thousands MT of vegetable 6988.47 thousands MT of spices and 16753.21 thousands MT of plantation crops, which is very low as compared with other country. This makes India to own the responsibility of leadership in generating basic information and protection technology for horticultural crops, because of increasing interest in industry, academia and health sciences for horticultural crops, the demand of horticultural crops are increasing manifold. Keeping in view of the above, a plan of strategy to breaking the yield barriers in horticultural crops are necessary. This book brings together the scattered knowledge on all aspects of diseases to vegetable, spices and plantation crops occurring in India. I would like to compliment Dr. R.S. Mishra for their sincere efforts in bringing out this publication "Diseases of vegetable, spices and plantation crops". I hope that the book will serve as an important resource materials for those engaged in task of improving and stabilizing horticultural production in the country.

Kumarganj

(V. P. Pandey)
Dean
College of Horticulture and Forestry
N.D. University of Agriculture & Technology,
Kumarganj, Faizabad, (UP),India

Preface

Plant disease has made a profound impact on the society. It is intimately linked to the well being of human kind. Agriculture, especially horticultural crops has all ready made considerable progress toward fulfil the requirement of vegetable and spices produce. Plant disease is a centre of developments in plant and microbes. Identification of microbes causing plant diseases enjoyed the attention of scientists for a long time. Description of disease symptoms, pathogen survival and their dispersal was included in this study. Most important knowledge about the effect of environmental factors such as high and low temperature, humidity, rainfall, sunshine and wind on the growth of pathogen have been included in the diseases. Vegetable crops *viz.* potato, tomato, cucurbit, cabbage and cauliflowers are suffering many classical pathogens which reduces quantities and qualitative yield. The emphasis on management of diseases has been recognized and includes the recent development in all aspect toward the integrated management with use of bio agent and new chemical molecules. Effort has also made to add new publication on the specific pathogens and diseases in this chapter.

India is the largest producer and exporter of spices in the world and known as the "Spices bowl of the world". There is increasing interest in industry, academia and health sciences for spices; the demand of spices is increasing manifold. Despite of tremendous scope of spices, we do not have a comprehensive publication on spices diseases and their management. This book brings together the currently scattered knowledge on all aspect of spices diseases occurring in India because the spice crops are attacked by many crucial pathogen causing diseases and influencing the productivity negatively. Diseases of plantation crops, nutritional and physiological disease have been written with gathering recent information; those are published in India and abroad. It is hope that book will be useful to students making their first acquaintance with this special field of plant pathology and it may serve as useful references for research workers. Omissions and mistakes are bound to be found in this book. I am ready to bear the full responsibility of making any misrepresentation of facts.

The author is indebted to many colleagues who have read one or more chapters of the manuscript. This includes Dr. L. P. Awasthi, Dr. S. P. Pathak. Dr. P. K. Singh, Dr. V. P. Pandey, Dr. Savita Gupta, Dr. K. P. Tiwari. Dr. Pramod Mishra and my son Mr. Arjit Mishra, daughter Miss Priyanshi. All the figures and drawings were taken from original field and slide.

Dr R S Mishra

Contents

Diseases of Spice Crops

Diseases of Plantation Crops

Chapter 1

Diagnosis of Plant Pathogens

The number of plant pathogens has grown on several thousand infecting plants. One of the major challenges for plant pathologists to develop new detection techniques for pathogen identifications that appropriate preventive measures can use for their management. Accurate diagnosis and identification of plant pathogens is a pre-requisite of disease management to sustain high yield potential of crops. The continuous efforts are being made to develop simple, reliable, rapid and safe method for disease diagnosis. Visual identification of plant disease is hard to perform by inexperience person and is limited particularly to disease affecting on aerial parts of the plants. It cannot be performed effectively in case of soil and seed borne diseases. Pathogens of several diseases are causing similar symptoms. Micro examination of diseased tissue for identification of pathogens on the basis of morphological characteristics is required specialist taxonomists. Taxonomist develops diagnostics tools for management of diseases which can be used for production of disease free planting materials. Develop to resistant plants either through classical breeding programme or through biotechnological tools for certification to disease free plants, diagnosis of plant pathogens involves clinical and etiological observations. While clinical observations include symptoms, transmission and distribution pattern, etiological observation includes examination with light and electron microscopes, serological and molecular detection. The indirect methods of detection become popular after development of detection methods based on serological procedures particularly, enzyme linked immunosorbent assay (ELISA). Introduction of recombinant DNA technology and development of techniques of genome-characterization, the prospect of utilizing nucleic acid become feasible for diagnosis of pathogens. Detection based on nucleic acid hybridization and PCR is more useful for pathogen like phytoplasma and fastidious bacteria, viroids, the naked viral RNA of low molecular weight and little genetic information is to be detected. The application of reverse PAGE and PCR made their detection possible. The genome of mung bean yellow mosaic virus (MYMV), tobacco leaf curl and potato virus XPVY, Potyvirus,water melon bud necrosis, toposvirus have been cloned and sequenced for identifying useful genes for developing virus resistant transgenic. The specific cDNA probes and RFLP analysis have helped for distinguishing strains of MYMV to prevent in the country. A multiplex PCR protocol has been standardised for these detection of RTV complex in vegetable.

Real Time PCR is using to detect the presence or absence of the target pathogen, and also quantify the amount present in the sample, because enumerating the pathogen upon detection is crucial to estimate the potential risks with respect to disease development and provides a useful basis for disease management decisions. Crops can be attacked by many pathogens; therefore many disease diagnostic applications required simultaneous detection and quantification of several pathogens. The DNA microarray technology, originally designed to study gene expression and generate single nucleotide polymorphism (SNP) profiles, is currently a new and emerging pathogen diagnostic technology.

Immunosorbent Assays in Plant Pathology

The past decade two immunosorbent procedures used into plant pathology have led to major revaluation of potential for detecting and identifying plant pathogens. The first is Serologicallly specific electron microscopy called Immunoserbent electron microscopy (ISEM), second is Enzyme linked immunosorbent assay (ELISA). It is precipitin based serodiagnostic techniques function by the polymeric interaction of specific antibody with homologous antigen to form immune complexes. These develop through the formation of antibody bridges into lattice- like structures which ultimately agglutinate (Particulate antigens) or form precipitates (soluble antigens), which is detected visually, because polymerization take place before the threshold of visual detection (Insolubility) is reached. Immunoprecipitation is a limited sensitivity for detecting antigens. Therefore to increase the sensitivity have generally followed one of two broad approaches:

1. The use of large, readily flocculated particles such as erythrocytes and betonies clay and polystyrene latex adheres as inert carriers for specific antibodies
2. The labelling of antibodies with directly detectable molecules *e.g.* fluorescent compounds or ferritin molecules.

Radioisotopes as labels are a highly sensitive system (RIA) for detection of antibody and antigen and significantly improved the potential for their quantitative assay. The use of enzymes in place of radioisotopes enabled much of the sensitivity. Different typed of enzyme immunoassay (EIA) are used in diagnostic. Homogeneous assay are used for low molecular weight example drugs, hormones *etc.* and heterogeneous assays are suitable for detection of plant and animal pathogens. Heterogeneous assays which are commonly used for detecting plant pathogens are double antibody sandwich procedure, first described in detail for plant viruses in 1977.

Principles

In the double antibody sandwich ELISA procedure the antigen (Pathogen) in the test sample is selectively trapped and immobilized by solid-phase specific antibody. Enzyme- labelled specific antibody is then reacted with the immobilized antigen and after removing unreacted enzyme-labelled antibody, the retained enzyme is assayed by adding a suitable substrate. The use of substrates that give colored hydrolysates permits easy visual qualitative assess ment of the presence or absence of pathogen in the sample. The quantitative measurement of the pathogen is possible by colorimetry.

In ISEM procedures pathogens (Antigens) are trapped in a similar way except that the solid phase is the microscopic grid. The immunospecifically trapped particulate pathogen is detected by direct visual observation.

Characteristics

1. The most significant characters of immunosorbent assays is their ability to detect plant pathogens occurring in very less concentration It is due to selective antibody for trapping specific pathogens and detection system is highly sensitive and discrimination to signal for the presence of the pathogen
2. In ISEM, detection is by direct visual observation of the pathogens
3. ELISA is its ability to detect antigens of very different size and morphology and same protocol can be used in all types of pathogens including fungi, bacteria and virus.
4. ELISA procedure provides quantitative data which is useful for pathologist and scientist for obtaining precise measurement of the occurrence and distribution of specific antigens and plant pathogens.

Procedure

Polystyrene microtiter plates have been used for the majority of immunosorbent assays in plant pathology. Plates now marketed for ELISA which are formulated and manufactured to ensure uniformly high protein adsorption capacity. For assaying limited number of samples polystyrene balls, cuvette strips or single well systems are available.

Antiserum

For ELISA test, antiserum quality is undoubtedly the most important, best results are obtained using high titer antisera, because suitability of an antiserum for use in ELISA should not be prejudged on the basis of precipitin titer alone. Some potentially suitable antisera seem to contain autoreacting components that impart to the purified Y-globulin preparations, an element of random reactivity or stickiness particularly noticeable when such antisera are used to prepare conjugates.

Antiserum Production

Rabbits are normally used to produce antisera to plant viruses and other plant pathogens. Young, immunologically competent animals (4-12 months) are most suitable as the serum obtained from older animals. The old animals may contain constituents that interfere with the reactions or that give rise to autonomous nonspecific reactions. Immune serum obtained too soon after injuction may be unsuitable for ELISA because of the predominance of low avidity IgM (19S) antibodies. Later bleeds will be proportionally richer in the more useful IgG (7S) antibodies. Broadly spectrum antisera can be produced by immunizing animals with virus subunits or protein or with mixed antigen preparations. Chicken are also used for used to provide serum or egg-yolk antibodies are used for second antibodies system for detecting plant viruses.

Y-globulin Purification

Partially purified preparations of Y-globulin may be obtained by any of the accepted procedures. Globulins are seprated from albumin by repeated salt precipitation at room temperature using eithera half saturated solution of ammonium sulphate or 18 per cent sodium sulphate followed by extensive dialysis against phosphate-buffered saline (PBS). Salt fractionated globulins can be used without further treatment. Partial purification of IgG can easily be achieved by filtering the preparation through a pad of diethylaminoethyl cellulose. This step effectively removes contaminating globulins, lipoproteins *etc.* and increase the activity and efficiency of subsequent conjugation procedures.

Choice of Enzyme and Conjugation Procedure

Two primcipal prerequisites for an effective enzyme- antibody conjugate are 1. That enzyme is stable and is detected easily and sensitively. 2 that a high proportion of the component enzymatic and immunological activity is retained in conjugate.

Alkaline phosphatise (ALP) and Horseradish peroxidise (HRP) have been the enzymes of choice in plant pathology. ALP gives essentially linear reaction kinetics with its substrate p-nitrophenyl phosphate over a protected period, enabling weak but specific reaction to develop with little or no increase in background colour. HRP conjugates are less flexible in this respect because of progressive inhibition of the enzyme by its hydrolysis products. Both enzymes can be coupled with Y-globulin in a one step procedure by mixing the reactants in the presence of dialdehyde, glutaraldehyde. This procedure gives high molecular weight conjugates well suited for detecting macromolecules such as plant pathogens. The concentration of glutaraldehyde (GTA) 0.015-0.06 per cent are used for the coupling reaction, influences the degree of polymerization. The efficiency of conjugation is rather low, about 3-8 per cent enzyme activity for ALP and 0.5-2 per cent for HRP.

Substrate

The substrate provides a sensitive means of detecting enzyme in the conjugate. Suibstrates, which are initially colourless but give a coloured hydrolysis product, are idial for application in plant pathology. The coloured product can be identified visually or quantified by colorimetry. For ALP conjugates p-nitrophenyl phosphate is the preferred substrate. For more sensitive reaction substrate methyl fluorescein monophasphate is used.

Antigen

Antigen is morphologically diverse as small proteins, nucleic acids, bacterial and fungal conidia have been assayed with a high degree of sensitivity and specificity.

Preparation of Sample for Antigen

Infected plant part used for antigen preparation. Some time many plant parts contain compounds which can inhibit the immunological or enzymetic reaction or give rise to undesirable nonspecific reactions. For this using additives incorporated in the extraction buffer to stabilize the antigen and to effectively neutralize compounds such as polyphenol oxidases and tannins. A PBS - Tween extraction buffer incorporating 2 per cent polyvinylpyrrolidone and 0.2 per cent bovine serum or egg albumen has been recommended for routine tissue extraction. Incubation plates should be kept at 25-30°C is less likely to produce thermal gradients than at 37°C. Overnight incubation of antigen at 4-6°C is usually is convenient for clean backgrounds. Coated plates may be stored after rinsing in plastic bags at 4°C for several weeks. For long term storage, plates should be dried and sealed in foil with a suitable desiccant.

Coating the Plates

The importance of optimizing the conditions for adsorbing Y-globulin to solid phase, for both ELISA and ISEM. For ELISA passive adsorption of Y-globulin is achieved either by applying a relatively high concentration of protein for short time (less than 30 minutes) or a lower concentration for loger time. Different quantitative results can be produced by these two approaches depending on the characteristics of the antigen to be detected and on the activity and specificity of the Y-globulin preparation.

Antiserum and Conjugate Factors

Antibody reacts with antigen by an association of relatively weak, shortrange forces which come into effect only if a high degree of steric complementarity exists between the two reactants. Thus the conjugates will not bind if the distribution of reactive antibody sites is incompatible with the distribution or accessibility of the antigenic determinants. Antibody and antigen interact the bond may be insufficiently strong to withstand the physical forces on the conjugate when the plates are rinsed.

Pathogen Diagnosis

Pathogen diagnosis with ELISA is done merely as positive or negative by visual observation of the presence or absence of colour.

Diseases of Vegetables

Chapter 2

Brinjal (*Solanum melongena*)

1. Phomosis Blight, Leaf Spot and Fruit Rot

The disease was first reported from Gujarat in 1914 and later to other parts of India. The causal pathogen causes seedling blight, leaf spots and fruit rot.

Symptoms

The nursery plants show damping off when plants are infected at seedling stage. Infected leaves show circular spots of cinnamon-buff colour with irregular blackish margin. Lesions may also develop on petioles and stem. The affected portion of the plant is blighted. On fruits, minute, sunken, dull and dusky spots appear first which coalesce later converting in rotten areas. Severely affected fruits show rotting of the entire flesh.

Pathogen

The disease is caused by *Phomosis vexans*. Numerous pycnidia are developed on the affected portion of the fruit coat. The fungus is specific pathogen for brinjal.

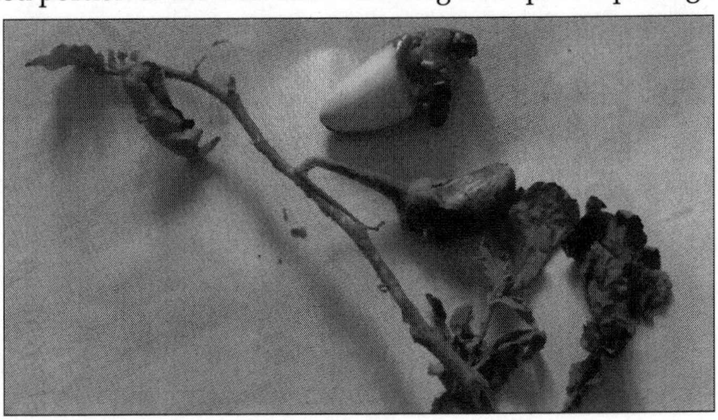

It is characterized by globose pycnidia with hard well-devepoped wall which is dark in colour. The perfect state is *Diaporth vexan*, which produced asci,the asci are cylindrical fusiform or clavate attached to ascogenous cells. They are producing 8 ascospores.

Disease Cycle

The pathogen survives in soil on affected plant debris and also externally seed-borne. When the seeds or seedling are planted in infested soil, the organism become active and infects the plants.

Control

Field sanitation and seed treatment with fungicides are important control measures. Treating the seeds with 0.1 per cent mercuric chloride or other organomercuric fungicides can eliminate fungal infection. In nurseries and fields, spraying with 1 per cent Bordeaux mixture helps in protecting plants from infection.

2. Bacterial Blight

The disease is found only in tropical and sub-tropical regions. In India, the disease has been reported from West Bengal. The pathogen also attacks several solanaceous crops such as potato, tomato and chillies and other cultivated plants which include castor, groundnut, banana, ginger and many wild plants. The disease is often associated with nematode infestation.

Symptoms

The most characteristic symptoms are wilting and stunting. The stem at the base becomes dark brown and constrict leading to wilting of plants.Sudden wilting and drooping of leaves occur when the infection is severe.

Pathogen

The disease is caused by the bacterium, *Pseudomonas solanacearum* (Smith) Smith. Three races of the pathogen have been reported and race-1 causes infection to solanaceous plants. Details of the pathogen have been mentioned under Bacterial wilt of potato.

Disease Cycle

The pathogen survives in soil or on alternate or cultivated or wild hosts. Soil is a potential source of primary infection. The disease destroys tomato, brinjal and chilli crops every year.

Control

Okra-cowpea-maize rotation reduces the infection by the bacterium.

Foliar application of agrimycin-100, chloramphenicol, or streptomycin sulphate is effective at 100ppm but prior to infection.

Rain or irrigation water should not be allowed to flow from sick field to healthy field.

Chapter 3

Elephant Foot Yam or Suran (*Amorphophallus paeonitolicus*)

1. Leaf Spots

Leaf spots of elephant foot yam are caused by many fungi such as *Alternaria amorphopholli, Cercospora amorphopholli, Corespora chevalieri.*

Symptoms

Spots aymptoms are very large query with brown margins, clusters of conidiophores are very dense. At lower leaf surface the spots are nearly coloured.

Pathogen

The mycelium is septate intercellular sending houstoria into palsade and mesophayll cells. The houstoria are branched. Conidiophores develop on a dense, globular, and brown to black stroma. They emerge by rupturing the epidermis in dense fascicles. Conidia are hyaline or pale yellow, obclavate 4-12 septate with round to distinctly truncate base and sub-acute tip.

Survival

The disease perpetuate through conidia lying in the soil and on diseased plant debris.

Dispersal

Soil and air are main source of pathogen dispersals.

Environment

24-28°C temperature, high relative humidity, prolonged low temperature and dew favour severe disease infection.

Control

1. Used healthy planting materials for disease free plants.
2. Spray the vines with dithane Z-78@ 2g/litre water in two to three times.
3. Spray the vines with cevcohim 2g/litre water in absence of above fungicide.

2. Amorphophallus Mosaic (Suran Mosaic Virus)

Casual Organisms

Amorphophallus Mosaic Virus

Symptoms

The characteristics symptoms of mosaic virus are mottling on the leaves. The diseases cause more proliferation on lateral buds, separation of buds from the mother corms and poor growth of roots. The infected plants produce small corms with low field.

Transmission

The virus is transmitted by aphids, *myzus persical* and *alphis qosspil.*

Control

1. Use corns from healthy plants only for seed purpose
2. Rogue out diseased plants as soon as first symptoms appeared

3. Collar Rot

Casual Organisms

Rhizoctonia solani Kuhn

Sclerotium rolfsii (Sacc)

Symptoms

The pathogen attacks the collar region and produces water socked lesions. The whole plant soon turns yellow; stem shrinks and collapses due to rotting. A thick white mycelium growth can be seen on the affected portion. Sclerotia (fruiting body) are also found on the mycelia growth.

Survival

Pathogen survive in soil and infected plant debris.

Dispersal

Pathogen dispersed through soil.

Control

1. Remove all plant debris after harvesting.
2. Drench the soil around the plant with capton solution (@2,0g/litre water) or brassicol solution (@1.5g/litre of water) twice at 25-30 days interval starting from first appearance of disease

Chapter 4

Okra or Ladies Fingers or *Bhindi* (*Abelomoschus esculentum* L.)

Leaf Spots

Cercospora leaf spots or Blight

Cercospora leaf spot caused by *Cercospora abelmaschi* Ell. and Ev, *Cercospora hibisci, C. hibicina* and *C. malayensis.*

Symptoms

The fungus grows as shooty mould on the lower surface of the leaves. Badly affected leaves roll, wilt and fall down. *Cercospora malayensis* causes brown, irregular spot with gray centre and dark red coloured margins. When the disease is severe, it produces complete defoliation

Pathogen

The mycelium is septate intercellular sending houstoria into palisade and mesophayll cells. The houstoria are branched. Conidiophores develop on a dense, globular, and brown to black stroma. They emerge by rupturing the epidermis in dense fascicles. Conidia are hyaline or pale yellow, obclavate 4-12 septate with round to distinctly truncate base and sub-acute tip. The perfect state is *Mycosphaerella,* they produce perithecial, it is ascostroma embedded in eaf tissues, broadly ovate to globose with papillate ostioles. Asci are cylindrical to clavate and contain 8 ascospores.

Disease Cycle

The fungi survive through the conidia and stomata on the plant debris in the soil. High moisture and high humidity favour the disease.

Other Leaf Spot Diseases

1. Phylosticta leaf spot caused by *Phyllostica hibiscini*
2. Alternaria leaf spot caused by *Alternaria hibiscinum*

Control

1. Spray the crops with copper oxychloride (fytolon or blitox-50)@3g/litre or dithane Z-78 or dithane M-45 or capton@2g/litre water at regular interval.
2. Treat the seed with organomercurials (0.2 per cent).

Yellow Vein Mosaic of Okra or *Bhindi*

The disease is caused by yellow vein mosaic virus (YVMV). This virus is transmitted by white fly (*Bemisia tobaci*) and the leaf hopper (*Empoaca devastone*). This virus has several wild hosts. This is the most devastating diseases of okra causing 80 per cent yield loss.

Symptoms

The characteristics symptoms are clearing of small vein and venial chlorosis of leaves. The yellow network of vein is very conspicuous and the vein and vein lets are thickend. The chlorosis extended to the interveinal areas and in case of severe infection the whole leaf turn yellow or cream colour. The infected plants remain stunted and bear very few deformed and small fruit, if plant gets infected within 20 days after germination. The fruits are dwarfed, molformed and yellowish green in colour. The infected fruits are unmarketable.

Transmission

The virus causing yellow vein mosaic of okra is transmitted in nature by white fly (*Bemisia tabaci*).

Epidemiology

Hot and dry weather favours multiplication of white fly population and YVMV incidence are observed during summer months (March-June)

Control

1. Apply thimet10G @20kg/ha or disyston granules @25-30kg/ha at the time of planting and often 40-50 days of first application.
2. The vector population and disease incidence can be reduced by application of 4-5 foliar sprays as either Dimethoate (0.05 per cent) or Metasystox (0.02 per cent) or Monocrotophos (0.05 per cent) followed by 1 or 2 minerol oil
3. Alternatively start spraying of dimecron and nuvan@1ml each (*i.e.* 2ml/3litre of water) just after 7-10 days of germination. Repeat spraying before 15 days of harvest 4-6 spraying are needed.
4. Adjustment of sowing time when population of white fly is least. Summer crops suffer more than rainy season crops.
5. Rouging of diseased plant.

6. Grow resistant variety *viz;* Parvani kranti, Arka abhay,Arka anamika,Punjab padmini, and IIHR Banglore.

Enation Leaf Curl of Okra

During the year 1983-84, a new disease namely enation leaf curl in okra was observed at IIHR, Banglore.

Symptoms

The initial symptoms of the disease are thickening of the veins followed by warty and rough texture of the leaves. Later on leaves began to curl in an adexial direction. Mild and bold enations are prominent on the under surface of the leaves. The most characteristic symptoms of this disease is twisting of the main stem and lateral branches. The leaf becomes thick and leathery. The leaf curling is more prominent in middle age leaf. Plant growth is retarded, fruit of infected plants are small and deformed. Fruits of infected plants are unfit for marketing.

Transmission

The virus causing enation leaf curl of okra is transmitted by both grafting and white fly (*Bemisia tabaci*). Inoculated plants exhibited typical symptoms after 20-25 days.

Control

1. One soil application of furadon @ 1.5 Kg a.i./ha at the time of sowing. Healthy seeds are recommended for sowing.
2. Infected plant must be removed
3. 4-5 spray of Momocrotophas (0.05 per cent) is recommended for control.

Powdery Mildew

Symptoms

White powdery growth on both sides of the leaves. In severe case diseased leaves drop off. These patches originate as minute discoloured specks from which a powdery mass radiates on all sides, when the infection become advanced large areas on the aerial parts of the host may be covered with these white floury patches. The superficial mass consists of mycelium and spores of fungus.

Causal Organism

The disease is caused by *Erysiphe polygoni* DC. Pathogen is obligate parasite and mycelium is fine, persistent, and rarely thick. Pathogen is ectophytic, entire thallus except houstoria are present on the surface of affected plants. The houstoria develop as outgrowth from lobed swellings (appresoria) on the side of hyphae adjacent to the leaf surface. The conidiophores arise vertically from the superficial hyphae on the host surface. Each conidiophore bears several spores in chain. The conidiophores are septate and their cells. Conidia are elliptical barrel shaped or sometimes cylindrical, hyaline, unicellular, 25-35x 13-16μm in size. The cleistothecia appear as sharp, black, minute bodies scattered in the mycelium net. They are round structures and composed of polygonal cells and are formed number of myceloid appendages. Generally 2-8 asci are formed in each cleistothecium. They are ovate, nearly sessile,46-72x 30-45µm in size. Each asci contain 3-8 ascospores which are elliptical hyaline, unicellular and measuring 19-25x 9-14µm.

Survival

Perenates through cleistothecia in soil.

Dispersal

Primary infection through ascospores and secondary infection through conidia by air.

Epidemiology

The pathogen survives on some weed hosts. Dry and warm weather favour the disease developments.

Control

Spray sulfex (0.2 per cent) or Karathane (0.05 per cent) at 10 days intervals.

Chapter 5

Pea (*Pisum sativum*)

Powdery Mildew of Pea

Powdery mildew disease is distributed throughout the world. The disease appears in epidemic form in every year when the plants are in pod stage toward the end of January and February.

Symptoms

Small irregular powdery spots appear on the upper surface of leave, when plants are in the flowering and pod stage during January and February. The powdery whitish spots completely cover the leaves, petioles, stem and even the pods. Plant appears greyish white appearance and leaves turn yellow. Yield may be greatly reduced.

Casual Organisms

Erysiphe polygoni

The mycelium consists of delicate and persisante hyphae which are attached to the leaves by appresoria, hyphae penetrate the epidermis and swell. Condiophore arise vertically from the leaf surface, conidia usually form single (rarely in chain), cleistothecia appear as sharp, black scattered on the surface of white mycelium, appendage very variable in number and length.

Survival

Fungus is obligate parasite, so fungus survives through cleistothecia in the soil. They are also seed borne.

Dispersal

Seed, soil and air

Epidemiology

The pathogen survives on affected seed and other weed hosts The disease spread in worm and dry weather. It is assumes epidemic proportions on pea (*Pisum sativum* L.) especially when lower temperature (less than 23°C) and relative humidity (40-70per cent) prevail even for short periods

Control

1. Field sanitation and disease plant refuse should be collect and burnt in the field
2. Early maturing variety should be used.
3. Dusting with sulphur @ 25-30 kg/acre.
4. 0.10per cent Karathane can be used for spray.
5. Seed treatment with organosynthetic fungicides should be used example capton and Cerenox.
6. Resistant variety T-10,P-185,P-388,P-6583

Rust of Pea

It is worldwide in distribution and also occurs in several part of India. It is a macrocyclic autoecious rust, first appears in the month of January.

Symptoms

Light yellow pustules are found on all above ground plant parts more on lower surface of leaves. Pustules become reddish brown in colour. Severely infected leaves shrivel and fall off. Stem infection results cracking and braking of stem which leads to drying and death of plants. Pod infection deteriorates the quality of fruits. Aecia are the earliest appear around in the month of February. Pycnia mostly hypophillous are in frequent. All the green parts including pods are affected. Aceia are hypophillous scattered over the entire leaf. Uredial sori develop both the side of leaf as other parts of plants. Teleuto pustule occurs in the same sorus as the uredia and develops from the same mycelium as the uredia. Symptoms are same in all parts of plants and shows black in colour.

Casual Organisms

Rust of pea is caused by *Uromyces fabae* (persoon) de bary. It is autoecious rust. Aceia spore are round to angular or broadly ellipsoidal 17-22μ diameters, yellow in colour. Wall hyline, finally verrucose. Urediospores broadly ellipsoidal 22-28x19-22μ thick. Teleuto spore are also ellipsoidal, ovoidal or cylindrical.

Survive

Aceia spore not survive but teleuto spore resist and survive for two year in high temperature. The pathogens are also survives on other host plants.

Environment

At 17-22°C temperature infection take place by the aecia while at 25°C temperature the infection caused by uredia. 3-18°C tempeature in the month of January out break take place. Cool and moderate temperature favour the disease mostly on winter in months.

Control

1. Use resistant variety.
2. Spray with zineb, ziram thiram, chlorani land thiocarbamoyal.
3. Agrosan and thiram can also used for seed treatment.
4. Early sowing minimized the disease infection.
5. Spray dithane M-45 (0.2 per cent) or sulfex (0.3 per cent) at 10 days intervals

Downy Mildew of Pea

Downy mildew is a severe disease on pea in areas where cool, footy weather prevails.

Symptoms

Disease appears on the lower sides of the leaf as fluffy, white to slightly blue, cottony patches with the fructification of pathogen. The areas of opposite sides turn yellow and may become brown and dry. The stem may also systematically infected and become distorted thus the growth of whole plant become stunted if weather found humid. Immature pods become infected and symptoms appear as brown blotches on the outside of the pods. Abundant oospores of the pathogen are formed on this tissue. Seeds adjacent to the proliferated areas of the pathogen aborted or showed small, brown, sunken spots.

Causal Organism

Peronospora pisi Syd.emend. Campbell.

The sporangiophores are dichotomously branched near the tip. They bear oval to elliptic, subhyline sporangia, 11-22 x 13- 29µm, which germinate directly by germ tube. At the letter stage coenocytic mycelium gives rise to oogonia and antheridia within the host tissue. After fertilization oogonia formed spherical, dark- walled, reticulate oospores formed which are 26-43µm in diameter. Oospore germination has not been observed.

Disease Cycle

Survival

Oospores are formed in the proliferated pods and mycelium occurs in seed coat. Oospores are adhered on the surface of the seed during threshing.

Dispersal

Oospores are also found in soil as well as seeds; the seedlings become infected, which provide primary sporangial inoculums. Secondary infection occurs through sporangia, sporangia are air borne. Penetration occurred by stomata. Mycelium are found intercellular except houstoria.

Environmental Condition

Cool moist weather favours epidemic development, whereas warm, dry weather retards the fungus growth.

Control

1. Crop rotation with other non leguminous crop should be follow
2. Frequent spray of wettable spergon on the crop
3. Spray Blitox or blue copper (0.4 per cent) or Dithane M-45 (0.2 per cent) or Difolatan (0.3 per cent) at 10 days intervals

Chapter 6

Crucifers

Leaf Spot of Crucifers (Cabbage and Cauliflower)

Alternaria leaf spot and pod blight of cauliflower although considered to be a disease of minor significance. In India have assumed seriousness in UP and Haryana. Early infection in young leaves causes severe foliage and stem blight leading to reduced flowering followed by infection on pods resulting in the production of srivelled and diseased seeds.

Symptoms

Leaf spot appear as small, dark coloured area which spread rapidly to form circular lesion upto 1cm diameter. The enlargement of spot formed concentric ring on the lower leaves. Bluesh growth of fungus appear on petiole, stem and leaf as linear spots. Head of cauliflower are also affected shows browning, beginning at the margin of individual flower on flower clusters. Fungus also affects scandling in the beds. It was seen that *A. brassicae* produced small, pale, brown to dark brown lesion

on cotyledons as well as a true leaves 144 hours after inoculation. Similar lesions also appeared on petioles. The lesion increased in size and progressively become more distinct with round or elliptical shape. Appearance of alternate black and brown rings and yellow haloes were most conspicuous in the expending lesions. *A.brassicicola* produced comparatively tiny and darker lesions and dark- black lesions increased slowly. Concentric rings were not conspicuous and yellow haloes were completely absent. The infected leaves curled, distorted and crumpled.

Alternaria brasscicola (*schw*) Wiltshire: Sunken and isolated black lesions, concentric rings are not conspicuous, yellow halo absent.

Alternaria brassicae (*berk*) Sacc: Pale brown to dark brown lesions, concentric rings conspicuous, spots surrounded by yellow haloes.

Casual Organism

Alternaria brasscicola (*schw*) Wiltshire *Alternaria-brassicae.* (*berk*) Sacc

Conidium of *A. brassicicola* germinated within 2hr producing 7-11 germ tubes and appresorial formation was seen after 12 hrs of inoculation. The infected hyphae penetrated directly as well as through stomata. Whereas, 2-5 small tubes are produced in *A. brassicae* after 5hrs. Hyphae of the fungus are branched septate mycelium; later brown, inter and intracellular conidiophore arise single or in group of 2-12 through stomata. Conidia are mostly in chain up to 20 or more. They

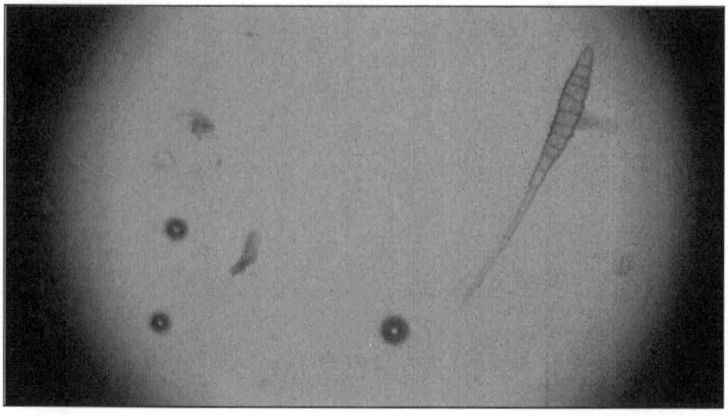

are straight, cylindrical, meuriform, usually tapering slightly toward the apex or obclavate and basal cell is rounded. Each conidium have 1-11 septa and less than six transverse septa and few or up to six longitudinal septa. Conidia are pole to dark brown.

Environment Condition

Pathogen favour hot moist weather, optimum temperature 20° C.

Survival

Pathogen survive on selected plant debris in an infected field.

Dispersal

Wind, insect *etc.*

Control

1. Seed treatment with chemical
2. Seed treatment with hot water (50° C for 30 minute)
3. Spraying of copper oxychloride (0.5 to 0.75 per cent) or maneb (0.2 per cent) or zineb (0.25 per cent) can be done as two to three times at 15 days interval.
4. Destroy the plant debris.

Yellow or Curly Blight

Cabbage yellows is a vascular disease

Symptoms

Disease affect to plant at any stage of seedling it may be affect at nursery. Normal symptoms appear 2-4week after transplanting, leafless yellowish green colour of the foliage and lateral curling of stem and leaves were observed in severe infected field. First lower leaf become yellow then upper leaf yellowed, tissue turn brown, brittle and dies. Affected plant become stunted, if cut the vascular bundle seen as yellow or brown. Smaller veins of disease leaf are much more.

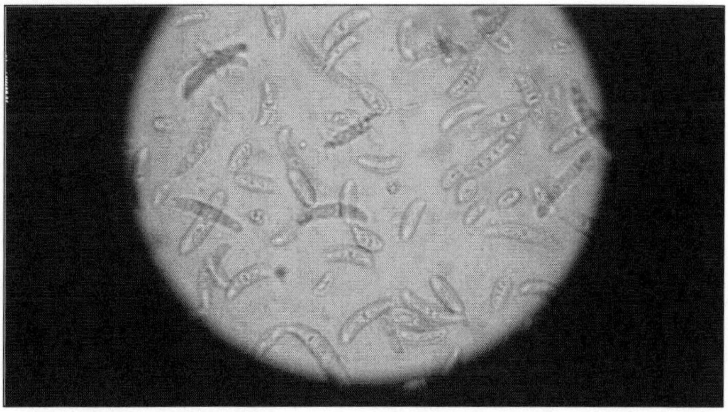

Causal Organism

Fusarium oxysporum f. Conglutinans

The fungus produce septate mycelium which is hyaline at first becoming cream coloured at age. Chlamydosphores are formed in older mycelium, single celled hyaline, ovoid to ellipsoid micro conidia are produce abundant, and macro conidia are rare.

Environment

Moist with warm weather favour the pathogen growth. The average soil temperature for mycelial growth should be 30 to 32°C.

Dispersal

Soil, wind and water

Survival

Plant debris and soil

Control

1. Sanitation,
2. Crop rotation,
3. Use of resistant variety,
4. Seed treatment with *Trichoderma* spp. @4.0g/kg seed

Downy Mildew of Cruciferous Crops

Symptoms

The yellowish, irregular or angular lesions with white downy growth appear on the stem, leaves and flowers. The symptoms on plants are similar to that caused by white rust. The curd tops turn brown. The stems develop dark brown depressed irregular streaks with greyish-whitish downy fungal growth occurs on under surface of leaves while corresponding upper surface have white spots. Infected leaves dry up shrivel and fall off. Flower port is deformed. The several infected curds become rot.

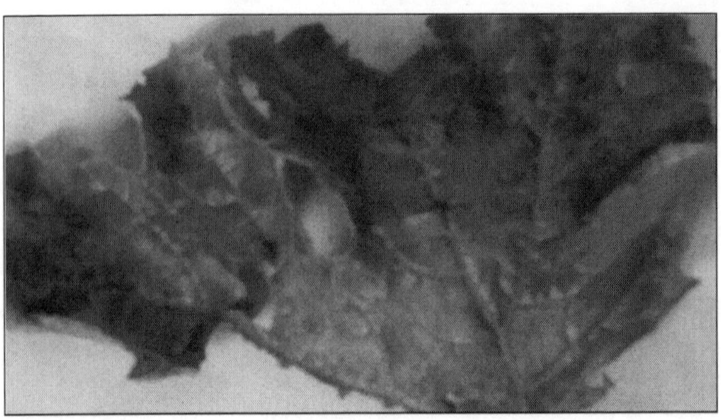

Causal Organism

Peronospora parasitica (Pers),ex Fr.

The fungus is obligate parasite; the mycelium is intercellular with large, finger-shaped or clavate, branched houstoria which nearly fill the cell cavity. After a period of vegetative growth numerous erect branched of sporangiophores emerge through stomata on the under surface of leaves. Only oospores are formed in floral organs. However sporangial production on the floral organs is profuse covering the surface with white downy growth. The sporangiophores are vertically and their base is flattened, these are 100-300μm long unbranched of major portion of their length. Dichotomous branching occurs at tip. The sterigmata are long slender and pointed. Single sporangium is borne on each tip. They are broadly oval, ellipsoidal, hyaline and measures 24-287 x 15-20μm and germinate by lateral germ tube. Oospores are produced in the interior of host tissue at latter stage of host. The oogonium is spherical and hyaline and fertilized by antheridium which produces oospores. They are globose crest like and appear pale yellow in colour. It is germinate by germ tube.

Epidemiology

The pathogen survives on diseased plant debris. Low temperature and high moisture favour the disease developments

Survival

Soil borne through oospores lying in diseased plant debris

Dispersal

Primary infection through soil and secondary infection occurs through wind

Control

1. Spray 0.2 per cent solution of Dithane M-45 at 7 days interval
2. Difolatan @ 0.3 per cent is effective for control of disease at 8-10 days intervals
3. Spray of Blitox/Blue copper (0.4 per cent) is found effective at 8-10 days interval
4. The oospores are inactivated in the soil if a rotation of two to three years is followed

White Rot or Stalk Rot (*Sclerotinia sclerotiorum*)

Symptoms

It is most serious disease of seed crop of cabbage and cauliflower. Stem and inflorescences twigs become straw and dry in March. Pith of infected plants is filled with black and hard sclerotia of fungus.

Control

Four to five spray of 0.05 per cent Bavistin in combination with 0.2 per cent Dithane M-45 followed by 0.2 per cent Dithane M-45 alone alternately at 10-15 days interval for the control of disease.

Club Root

Records of this disease go back as far as the thirteenth century in Western Europe, but first intensive report of these diseases was carried out by Woronin in 1874. Clubroot is now known in many parts of world.

Causal Organism

Plasmodiophora brassicae Wor.

The resting spore is hyaline, spherical upto 4µm in diameter. It germinates in a favourable medium and atmospheric condition without a prolonged resting period, swelling to several times with the original size and giving rise to a single anteriorly biflagellate primary zoospore. The zoospore is naked, uninucleate protoplast is usually exceedingly active by irregular jerks for several minutes and assuming an amoeboid motion. Young amoebae are found in the root hairs and cortical cells of roots of cabbage. The zoospores of saprophytic slime moulds fuse before the development of plasmodium. The plasmodium is always found within a host cell. Letter stage cleavage take place around each nucleus in the plasmodium and separate in the resting spores. The spores are free from one another.

Symptoms

Root enlarge to form spindle shaped or spheroid growth; in-vasion by secondary organisms causes rotting and blackening of malformed tissues Galls are formed on the affected roots giving a club like appearance. Leaves of infected plants turn yellow and finally death of plant. It is prevalent seriously on the acidic soils having pH below 7.0 and less serious on heavy soils and those contain low organic matter.

Disease Cycles

Infected subterranean plant parts disintegrate in the soil and releasing the free spores, few spores germinate and large number of spore goes to long period of dormant stage or many years resting periods. Yet soils are known to remain infested for 10 years without the presence of host plants. Infection occurred at soil temperature of 9 to 30°C and found limited infection at low soil moisture. If moisture content of dry infested soil was raised to the level of optimum temperature 18-24°C, germination and penetration followed and disease development occurred. The organism is disseminated locally with drainage water, farm implements, wind-borne soil, moving animals. It is not seed borne disease.

Control

1. Fallow crop rotation
2. Raise the soil pH above 7.0 with liming.
3. Dip the roots in 0.05 per cent solution of PCNB at the time of transplanting.

4. Treat seedlings at time of transplanting with Hg Cl_2 solution (1:15000 at the rate of. 125ml/100 sedling

Black Leg

Phoma lingam

Black sap stream can be seen in the affected plant stem when split vertically. Light brown lesions are formed on the stem and finally plant wilts. It is a seed borne disease.

Control

1. Grow resistant varieties like Pusa Drum Head of cabbage.
2. Follow 3 years crop rotation with non-cruciferous crops.
3. Hot water treatment of seeds at 50°C for 30 minutes

Bacterial Diseases

Black Rot or Black Spot

First infection occurred through water pores at the margin of leaves and first sign of disease appeared at this point. Blighting of leaves from margin to mid- rib in "V" shape and blacking of vascular bundles is characteristic symptoms; it is an internally seed borne disease. Foliage of affected plants becomes yellow followed by blackening of veins and death of whole plant. Veins and vein lets become dark in colour. Vascular discoloration extended to main stem and proceeds up and down wards. Abscission layers are formed at the base and affected leaves may develop prematurely and early defoliation occurs. Cabbage and cauliflower heads are invaded and discoloured. When pathogen attack is early and severe then head/curd formation is restricted.

Causal Organism

Xanthomonas compestrisi (Pam.) Dows

The organism is rod shaped 0.7-3.0 x 0.4-0.5µm with single polar flagellum.

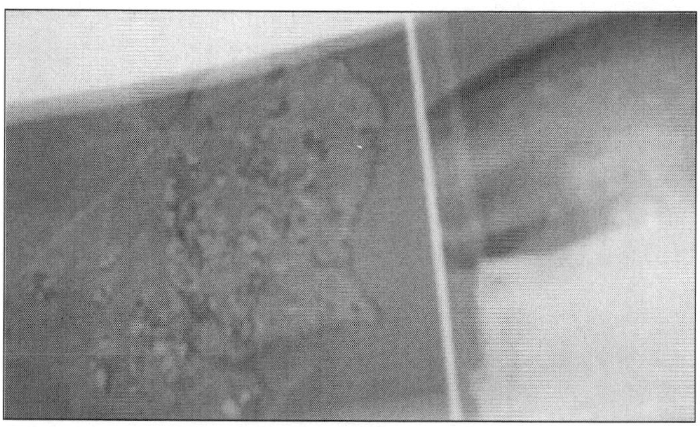

Survival and Spread

Pathogen persists in soil, seed and plant debris. It is seed borne disease. Bacteria mostly enter by rain drops in leaves through natural openings or wound or stomata as marginal sinus of the cotyledons, to which the infected seed coat is often attached and carried out above ground parts during growth. Bacteria are disseminated by wind-blown rain and cultivated implements. Seeds and transplant materials are wide responsible for spread.

Epidemiology

Disease become more serious during rainy season.

Control

1. Grow resistant varieties
2. Follow crop rotation with non cruciferous crops
3. Give hot water treatment to seeds at 50°C temperature for 30 minutes to eliminates both externally and internally seed borne bacteria.
4. Give the seeds treatments with 0.01 per cent Agrimycin or steptocycline or 0.1 per cent mercuric chloride for 30 minutes, which eliminates externally seed borne bacteria.

Soft Rot

Erwinia carotovora

Symptoms

Soft water soaked lesions seen on roots. It is characterized by a soft, slimy and bed smelling rot of affected plants. High atmospheric humidity favours the disease. The bacterial attack occurs only after the black rot or after mechanical or frost injury.

Survive

Pathogen survive on disease plant parts.

Control

Spray the crop with streptocycline or plantomycine @100-200ppm combine with 0.3 per cent copper oxychloride. Repeat the spray at 15 days interval.

Chapter 7

Carrot (*Daucus carota*)

Leaf Spot

Pathogen

Cercospora apii. The mycelium is septate intercellular sending houstoria into palisade and mesophayll cells. The houstoria are branched. Conidiophores develop on a dense, globular, and brown to black stroma. They emerge by rupturing the epidermis in dense fascicles. Conidia are hyaline or pale yellow, obclavate 4-12 septate with round to distinctly truncate base and sub-acute tip. The perfect state is *Mycosphaerella,* they produce perithecial, it is ascostroma embedded in eaf tissues, broadly ovate to globose with papillate ostioles. Asci are cylindrical to clavate and contain 8 ascospores.

Survival and Spread

Survive on infected plant debris and spread by air borne conidia

Symptoms

Circular yellowish coloured spots becoming necrotic grey with age.

Control

Foliar spray with 0.2per cent Dithane M-45

Leaf Blight

Alternaria radicini

Survival and Spread

Plant debris and spread by air borne conidia

Symptoms

Irregular spots with yellowish haloes. Blighted leaf with irregular spread of infection.

Control

Foliar spray with 0.25 per cent of Dithane M-45

Powdery Mildew

Erysiphe polygoni

Symptoms

Powdery growth seen on leaves and petioles. In later stage leaves turn in brown colour and lostly plant wilted. Pathogen reduces yield in considerable amount.

Sweet Potato (*Ipomoea batatus*)

Leaf Spot

Cercospora batatae, C. batattica

Symptoms

Large dark brown spots; severe damage to leaves in warm moist weather.

Survival

Infected plant debris and spread by air borne conidia.

Control

Foliar spray with sulfex (0.3 per cent) at fortnight interval.

Tuber Rot, Charcoal Rot, Black Rot, Soft Rot

Pathogen

Macrophomina phaseolina, Rhizoctonia solani

Symptoms

Slow storage rot is caused more damage when tuber are stored under high humidity.

Control

1. Raise vines from healthy locality,
2. Keep walls and floor of storage dry and disinfected,
3. Treat seed tuber with 0.1 per cent mercuric chloride before sowing.

Chapter 8

Cucumber (*Cucumis sativus*)

Leaf Spot

During 1931, a severe leaf spot disease on cucumber (*Cucumis sativus* L.) was observed in Pusa Bihar and 1983 in Sabour. The disease manifested as minute water soaked flecks on the abaxial surface and yellowing on the corresponding adaxial surface of leaf, the flecks enlarged and become angular on ageing. Transverse section of the leaf spot exhibited myriads of bacteria oozing out from the cut surface.

Causal Organism

Xanthomonas compestris pv. *Cucurbitae* (Bryan) Dye.

A yellow bacterium was isolated from the lesion, which inoculated on the host plants, produced water soaked spots 96-120 hours after inoculation. The colour of spots later turned to brown. The bacterium was rod shaped measuring 1.0 -1.4μm × 0.4 -0.6μm. Gram negative, non spore forming, polar monotrichous and aerobic. It produced yellow, raised and slime with entire margin of colonies on nutrients agar.

Control

1. Plantomycin @ 0.01 per cent are a good antibiotic followed by Paushamycin @ 0.01 per cent and streptomycine @ 0.015 for control of leaf spot disease of cucumber
2. Moderate resistant cultivars are Japanese long green, collection 72-10

Diseases of Cucurbits

Fusarium Wilt of Cucurbits

This is an important disease of cucurbits and found throughout the world and several species of *Fusarium* are known to cause the diseases in plants of family cucurbitaceous.

Symptoms

In young seedlings and cotyledons are dropped and weather due to wilt disease. Older plants wilt suddenly and vascular bundles at the collar region show brown discoloration.

Pathogen

The fungus, *Fusaium oxysporvum* Schl. is the causal pathogen. The pathogen belongs to family Tuberculariaceae, order Moniliales and sub-division Deuteromycotina. The pathogen produces both macro and micro conidia. The conidia germinate by germ tubes and produce hyphae and mycelium.

Disease Cycle

This wilt is a soil borne disease. Inoculums increases more rapidly in sandy soils than in clay, silty or loam. The inoculums can reach to high level after just two successive crops. The fungus available in soil penetrates the root system through injury. After infection the fungus multiplies and moves to xylem vessels which are later plugged by the pathogen mycelium. The blocking of xylem vessels interrupts the movement of nutrients resulting wilting of plants.

Control

Crop rotation with resistant varieties against fungus is viable methods of overcoming the disease. Soil drenching with fungicides like carbendazim *etc.* can reduce the inoculums. Soil solarization for 2-4 weeks by using impermeable plastics after amending the soil with ammonium sulphate is a good method to control the disease. Growing resistant varieties is the best method.

Downy Mildew of Cucurbits

This is an important disease of cucurbits such as sponge gourd, ridge gourd, bottle gourd, bitter gourd, snake gourd, pumpkin and cucumber *etc.*

Symptoms

Leaves first show mosaic like mottling. The pale green areas are separated by dark green islands. Soon the angular yellow coloured spots restricted along veins develop on the upper surface of leaves. Below the spots on the lower side of leaf purplish downy growth appears. The affected leaves die quickly. Usually middle leaves are infected first followed by other leaves. Only a few small fruits with poor taste develop on diseased plants.

Pathogen

The fungus *Pseudoperonospora cubensis* (Berkeley and Curtis) Tostowzew is the causal pathogen. The pathogen is an obligate parasite. The sporangiophores arise in a group of 1-5 from the intercellular mycelium. Sporangia develop on the branches of sporangiophores usually during night and are dispersed during morning hours. The sporangia produce zoospores upon germination which cause the infection to plants. Oospores formation is not common in this species of fungus but has been reported from Madhya Pradesh, Punjab and Rajasthan.

Disease Cycle

The pathogen infects large number of wild hosts from where the primary inoculums come to cultivated crops. In areas where oospores are formed, the primary infection may be through them. The disease in the field spread through sporangia which are disseminated by splashing rains and beetles.

Cucumber Mosaic

Symptoms

The disease is caused by cucumber mosaic virus (CMV). The important symptoms of the disease are mosaic, mottling, stunting and distortion of leaves. Fruits from infected plants exhibited yellowish green mottling or yellow and green spots and wart-like projections.

Transmission

The disease is transmitted through aphids.

Pathogen

The disease is caused by a virus of cucumovirus group in the family Bromoviridae. The virus particles are isometric, 29 mm in diameter. Genome of the virus is single stranded RNA and consists of 18 per cent RNA and 82 per cent protein. The virus is transmitted by large number of aphid species such as *Myzus persicae, Aphis craccivora etc.* in a nonpersistent way. It is transmitted through seeds of 19 plant species.

Disease Cycle

The virus inoculum is present throughout the year on one crop or the other or on weeds from where it is brought to new crop by insect vectors. Contaminated seed is another source of primary inoculums.

Control

Control of cucumber mosaic virus is very difficult. However, virus resistant cultivars or transgenic plants can give protection from the virus. Growing cucumber in polyhouses is advantageous as the crop remain free from virus resulting in more yield and quality of fruits.

1. It is advised that all weed should be removed from the field
2. Infected plants should be removed as soon as visible.
3. Spray of Dimecron (0.05 per cent) or Metasystox (0.02 per cent) at 10 days intervals.

Anthracnose Disease

Symptoms

The spots on the leaves are irregular, angular to circular water soaked, yellowish brown and finally become brown in colour. The leaf spots dehydrate, shrink, dry out at the centre and make irregular spot holes. The spots on the fruits are circular, water soaked- depressed and dark broun colour.

Causal Organisms

Colletotrichum gloesporoidia

The myceliums are narrow, sparsely- septate hyphae which are at first hyaline but later take on slightly dark colour. Aservuli are formed abundantly on the affected surface. They develop at first as tangled sub-epidermal masses of hyphae which arise numerous closely packed conidiophores which partially arise in the epidermis. Each conidiophores bear the conidium at apex setae are common. The conidium remains embedded in a viscid fluid which swells in moist condition. Rupture the epidermis and exposes the conidial mosses and disseminated by rain drop and insect.

Disease Cycle

Pathogen survives in infected seed, plant debris and other colletral host of cucurbit family.

Spread

Air, wind and insect.

Environments

Moderate high temperature and high rain fall favour the disease development

Control

Seed treatment with Bavistin (0.2 per cent) or Benlate (0.2 per cent). Spray dithane M-45 (0.25 per cent) at 8 -10 days intervals Spraying of fungicides, zineb, mancozeb and tricop-50 has been recommended to control the disease. Removal of infected vines and eradication of wild cucurbits from near the vegetable growing areas are economic control measures in view of the cash value of the crop.

Powdery Mildew of Cucurbits

This is a serious disease of cucurbits especially on pumpkins and bottle gourd.

Symptoms

White to dirty grey powdery spots appears on leaves and stem. These spots enlarge in size and can cover the whole surface of the host. Leaves are fall down during severe infection. Fruits on affected plants are smaller in size. Humid condition favours the fast development of disease.

Pathogen

Two fungi, *Erysiphe cichoracearum* D.C. and *Sphaerotheca fuliginea* (Schlecht.) Poll. He has been reported as pathogens of powdery mildew of cucurbits.

The size of conidia of *E. cichoracearum* varied according to physiologic races and Cleistothecia are not common. The number of ascospores is usually two, rarely three in each ascus. The ascospores are oval or sub cylindrical with variable size The fungus is heterothalic and obligate parasite on host.

S. fuliginea is known to cause powdery mildew of cucurbits in India. The perithecia are formed from September to February and each perithecium has one ascus which is broadly elliptical to subglobose. Ascospores are single celled, hyaline spherical, 6-10 ascospores per ascus. Cleistothecia are formed which can survive from one season to next season and infect to the crop.

Disease Cycle

Perithecia develop on residue of infected plants left after the harvest of cucurbits. In isolated areas, it may serve as a source of primary inoculum. The conidia from wild cucurbits may be another source of primary inoculum. The conidia are dispersed by wind or insects and infect the plant through epidermal cells of the host.

Control

Field sanitation and dusting with 200-mesh sulphur @ 25-30 kg/ha is quite effective to control the disease. Sprays with other fungicides like calexin (0.1 or 0.05 per cent), copper sulphate and karathane are also effective. Resistant varieties of each crop should be grown.

Chapter 9

Chilli (*Capsicum annuum*)

Chilli diseases caused a serious problem during the cultivation of pepper. There are three types of diseases: viral, bacterial and fungal infections. Try to know and recognize the symptoms are definitely the first step to make the right cure. the best cure is still prevention.

Damping Off

Symptoms

Seedlings killed before emergence. Water soaking and shriveling of stem. Factors favouring infection of pathogens are moist soils, poor drainage, 90-100 per cent R.H and soil temperature 20°C.

Pathogen

Pythium aphanidermatum.

Mycelium is hyaline, coenocytic, zoosporangia are lobed and branched. Zoospores are biflagellate and oogonia are spherical with smooth walled.Antheridia are monoclinous, intercalary or terminal. Oospores are aplerotic, single with thick wall.

Mode of Spread and Survival

The pathogen is soil borne. Zoospores spread through irrigation water. The disease spreads to main field by infected seedlings.

Management

Soil drenching with copper oxychloride 0.25 per cent.

Fruit Rot and Die Back

Symptoms

As the fungus causes necrosis of tender twigs from the tip to backwards of branches such disease is called die-back Infection usually begins when the crop is in flowering stage. Flowers drop and dry up. There is profuse shedding of flowers. The flower stalk shrivel and dry up. This drying spreads from the flower stalks to the stem and subsequently causes die-back of the branches and stem. Partially affected plants bear fruits which are few and of low quality. On the surface of the soil, the necrotic areas are found separated from the healthy area by a dark brown to black band.

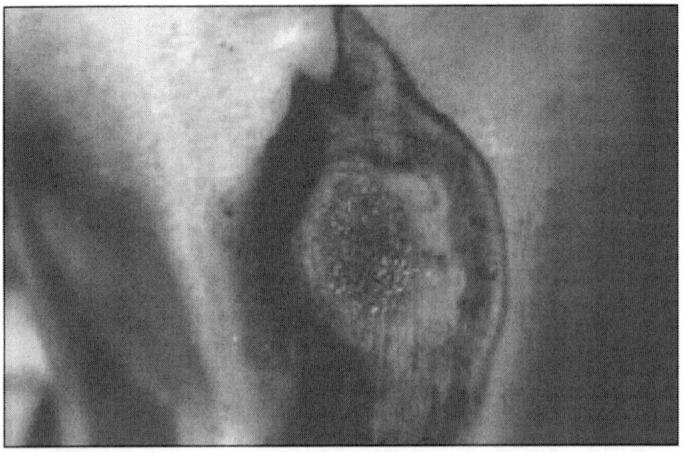

Pathogen

Colletotrichum capsici

The mycelium is septate, inter and intra cellular. Conidia in mass, appear pinkish. They are borne singly at the tip of conidiophores.

Mode of Spread and Survival

The fungus is seed borne and secondary infection is by air borne conidia and

rain. The disease spreads rapidly by wind blown rains during rainy season. Flies and other insects are found responsible for dissemination of the spores from one fruit to another. The fungus may not survive long in soil, but may survive on the dead twigs stored under dry conditions. Seeds from badly diseased fruits may also carry the primary inoculum.

Management

Use of disease-free seeds is important in preventing the disease. Seed treatment with Thiram or Captan 4g/kg is found to be -effective in eliminating the seed-borne inoculum. Good control of the disease has been reported by three sprayings with Ziram O. 25 per cent, Captan 0.2 per cent or Blitox-50 0.2 per cent. Chemicals like wettable sulphur 0.2 per cent, copper oxychloride 0.25 per cent and Zineb 0.15 per cent not only reduced the disease incidence but also increased the yield of fruits. The first spraying should be given just before flowering and the second at the time of fruit formation. Third spraying may be given a fortnight after second spraying.

Powdery Mildew

Symptoms

Shedding of foliage. White powdery growth on upper side of leaves.

Pathogen

Leveillula taurica

Disease Cycle

The powdery mildew disease cycle (life cycle) starts when spores (known as conidia) land on a chilli leaf. Spores germinate much like a seed and begin to grow into the leaf. chilli powdery mildew parasitizes the plant using it as a food source. The fungus initially grows unseen within the leaf for a latency period of 18-21 days. Then the fungus grows out by the breathing pores (stomates) on the under surface of the leaf, producing spores which are borne singly on numerous, fine strands or

stalks (conidiophores). These fungal strands become visible as white patches or mildew colonies on the under side of the leaf. Repeated cycles of powdery mildew can lead to severe outbreaks of powdery mildew that economically damage the crop.

Management

Spray Wettable sulphur 0.25 per cent or Dinocap (Karathane) 0.05 per cent.

Bacterial Leaf Spot

Symptoms

The leaves exhibit small circular or irregular, dark brown or black greasy spots. As the spots enlarge in size, the centre becomes lighter Surrounded by a dark band of tissue. The spot coalesce to form irregular lesions. Severely affected leaves become chlorotic and fall off. Petioles and stems are also affected. Stem infection leads to formation of cankerous growth and wilting of branches. On the fruits round, raised water soaked spots with a pale yellow border are produced. The spots turn brown

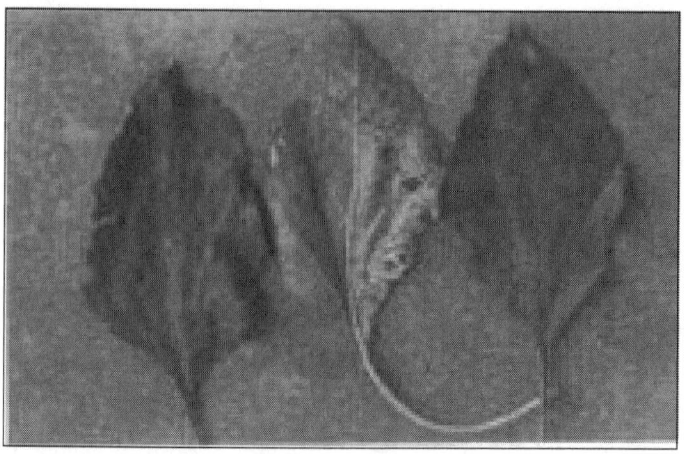

developing a depression in the centre, where in shining droplets of bacterial ooze may be observed.

Pathogen

Xanthomonas campestris pv. vesicatoria

Mode of Spread and Survival

The disease is primarily seed borne. It spreads in the nursery and is further disseminated through infected transplanting. Spattering rains are the chief means of dissemination. The bacterium subsists in infected debris.

Management

Seed treatment is found effective with 0.1 per cent mercuric chloride solution for 2 to 5 minutes. Seedlings may be sprayed with Bordeaux mixture 1. Per cent or copper oxychloride 0.25 per cent. Spraying with streptomycin should not be done after fruits begin to form. Field sanitation is important. Seeds must be obtained from disease free plants.

Cercospora Leaf Spot

Symptoms

Leaf lesions typically are brown and circular with small to large light grey centers and dark brown margins. The lesions may enlarge to 1cm or more in diameter and some times coalesce. Stem, petiole and pod lesions also have light grey centers with dark borders, but they are typically elliptical. Severely infected leaves drop off prematurely resulting in reduced yield.

Pathogen

Cercospora capsici

Stromata are well developed. Conidiophores are 30-60 x 4.5-5.5µm. Conidia are subhyaline to coloured, acicular to obculate.

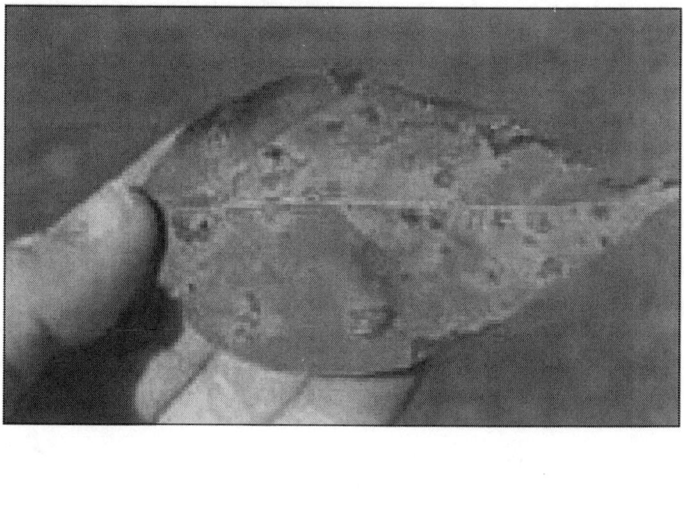

Mode of Spread and Survival

Primary source of infection are infected seeds, volunteer plants and infected plant debris. Secondary spread is through air borne conidia.

Management

Spray twice at 10-15 days interval with Mancozeb 0.25 per cent or Chlorothalonil (Kavach) 0.1 per cent.

Fusarium Wilt

Symptoms: Fusarium wilt is characterised by wilting of the plant and upward and inward rolling of the leaves. The leaves turn yellow and die. Generally appear localised areas of the field where a high percentage of the plants wilt and die, although scattered wilted plants may also occur. Disease symptoms are characterised by an initial slight yellowing of the foliage and wilting of the upper leaves that progress in a few days into a permanent wilt with the leaves still attached. By the time above - ground symptoms are evident, the vascular system of the plant is discoloured, particularly in the lower stem and roots.

Pathogen

Fusarium oxysporlum f.sp.capsici

Mycelium is grayish white. Microconidia are formed singly, hyaline and cylindrical. Macro conidia are cylindrical to falcate. Chlamydospores are globose to oval and rough walled.

Management

Use of wilt resistant varieties. Drenching with 1 per cent Bordeaux mixture or Blue copper or Fytolan 0.25 per cent may give protection.· Seed treatment with 4g *Trichoderma viride* formulation or 2g Carbendazim per kg seed is effective. Mix 2kg T.viride formulation with 50kg FYM.

Sprinkle water and cover with a thin polythene sheet, when mycelia growth is visible on the heap after 15 days, apply the mixture in rows of chilli in an area of one acre.

Leaf Curl

Leaf curl occurs toward the midrib and become deformed. Stunted plant growth due to shortened internodes and leaves greatly reduced in size. Flower buds abcise before attaining full size and anthers do not contain pollen grains. The virus is generally transmitted by whitefly. So control measures of whitefly in this regard would be helpful.

Mosaic Viruses

Light green and dark green patches seen on the leaves. Stunted plant growth was observed during early stages of crop. Yellowing and chlorotic ring spots on leaves and fruits are also recorded.

Management of Viral Diseases

Control measures are not known for majority of viral diseases. Hence, mechanical, cultural methods are mostly recommended. The infected plants should be uprooted and burnt or buried to avoid further infection. Avoid monoculture of chilli crop. Selection of healthy and disease free seed. Suitable insecticidal sprays reduce the incidence of viral diseases, since majority of viral diseases are transmitted by insect vectors. Soaking seeds in a solution containing 150 g Trisodium orthriphosphate per litre of water for 30 minutes inhibits seed-borne inoculum.

Treated seed should be washed with fresh water and dried before sowing. Nursery beds should be covered with nylon net or straw to protect the seedlings from viral infection. Raise 2-3 rows of maize or sorghum as border crop to restrict the spread of aphid vectors. Apply Carbofuran 3G @ 4-5 kg/acre in the mainfield to control sucking complex and insect vectors selectively. If it is not possible spray

the crop with systemic insecticides. Like Monocrotophos 1.5 ml or Dimethoate 2ml of Acephate 1g per litre of water. Collect and destroy infected virus plants as soon as they are noticed.

Bacterial Soft Rot

Symptoms

The fleshy fruit peduncle is highly susceptible and infections are frequently occurs at the initial point of fruits. Both ripe and green fruit may be affected. Initially, the lesions on the fruit are light to dark-colored, water-soaked, and somewhat sunken. The affected areas expand very rapidly, particularly under high temperatures, and tissues lose their texture. In later stages, bacterial ooze may develop from affected areas, and secondary organisms follow, often invading the rotted tissue. Post-harvest softening of stem and fruit end. The affected fruit hang from the plant like a water-filled bag.

Pathogen

Erwinia carotovora subsp. *Carotovora*

Disease Development

The bacteria may persist in fields where peppers are rotated with other susceptible crops such as cabbage and potato. The bacteria may be present as a contaminant on the surface of pepper seed. The bacteria can be transmitted by drainage water, irrigation water, or by sprinkler irrigation, but a wound is necessary for infection to occur. Wounding often arises from rough handling of plants during weeding, or due to a strong wind, or from insect feeding. European and Asiatic corn borers may introduce bacteria into the fruit peduncle of pepper during feeding. A high rate of nitrogen fertilization is associated with increased susceptibility to soft rot. Warm, moist weather is also highly favorable for infection.

Management

Use chlorinated wash water to reduce populations of soft rot bacteria and to reduce the risk of infection during washing. This will not reduce soft rot development in fruit infected with the bacterium prior to harvest. Allow fruit to dry thoroughly. During packing and storage, the fruit should be kept clean and maintained in a cool, dry place.

Alternaria Rot

Symptoms

The fungus is reported to enter wounds (sunscald or punctures). Dusty black spores on fruit spots are characteristic. In most instances this disease follows blossom-end rot, but it also follows injuries, chilling, and other decays. On the fruit, large greenish-brown to brown lesions covered, with grayish-brown mold are produced. Similar lesions on the lower-part of the fruit are characteristic of *Alternaria* rot following blossom-end rot. The larger lesions may show alternating light and dark-brown concentric zones. Shipping peppers under standard refrigeration will check the development of this rot, but when the fruit is removed from refrigeration the decay will advance rapidly at moderate to warm temperatures.

Pathogen

Alternaria sp.

Hyphae are septate, branched, light brown becoming darker with age and inter and intra cellular. Conidiophores emerge through stomata. Conidia are single and muriform.

Mode of Spread and Survival

Infected seeds, volunteer plants and infected plant debris are primary source of infection.

Management

The effectiveness of a pre-storage dry heat treatment and hot water dip in reducing storage rots of capsicum caused by *Alternaria alternata*. Treatment with hot air at 38°C for 48-72 h or hot water at 50°C to 53°C for 2 to 3 min, resulted in reduction in the pathogencity and development of these pathogens in inoculate peppers.

Chapter 10

Potato (*Solanum tuberosum*)

Late Blight

Causal Organism

Phytophthora infestans

Although effective measures for control exist, late blight remains the most serious fungus problem in many potato production regions.

Symptoms

Water-soaked lesions appear on foliage which, within a few days, turns brown then dry or black when weather condition found favourable such as cloudy weather with low minimum temperature and high relative humidity. Under moist conditions, white mildew-like sporulation is visible, especially on the lower surface of leaves. A Pale yellow margin often forms around leaf lesions). Lesions on stems and petioles are black or brown and the whole plant may collapse within a few days.

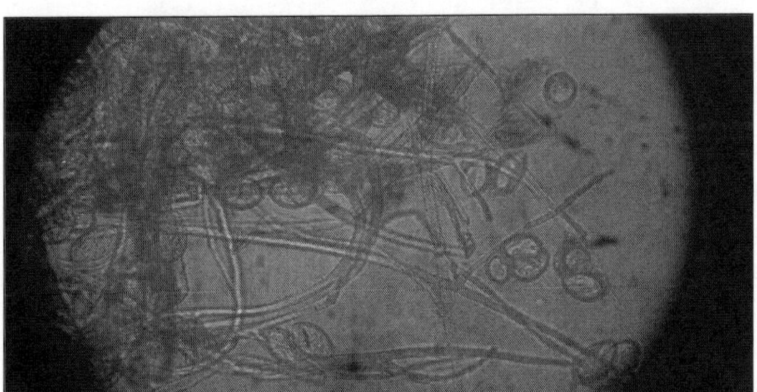

Potato tubers are infected with spores reached from the leaves and stems into the soil through rain washed. Initial symptoms on tubers have found brownish surface with discoloration. Sections cut through tubers show brown, necrotic tissues not clearly differentiated from the healthy portions. Later secondary rot develop in blighted tissues and this rots spread in storage.

Disease favoured by temperatures between 10°C and 25°C, accomplished by heavy dew or rain.

Survival

Fungus does not survive in soil; they survive only in plant tissue. Source of inoculums are neighbouring field of potato or tomato, volunteer plants.

Control

1. Use disease resistant cultivar
2. Prevent tuber infection through well chilled plants, healthy plants to be permitted for maturity and infected plants should be die naturally or killed before harvest
3. Three spray of fungicides *viz.,* mancozeb@2.5g/litre of water or ridomil@2.0g/litre or corbendazim/bavistin @2.5g/litre water should be done at 15 days interval.

Powdery Mildew

Powdery mildew is widely distributed on many host plants and develops in severity on potato under arid condition with high humidity.

Symptoms

Old infections superficially resemble late blight as leaves turn black, die and drop from the plant.Stem may also be infected, initially infected leaves are covered with whitish spore masses superficially resembling soil residues, dust or spray deposits.

Causal Organisms

Erysiphye cichoracearum

Environmental Condition

The disease requires high humidity but seldom develops where water falls on the leaves as rain or sprinkler irrigation.

Control

Control with elemental sulphur dust or fungicide sprays.

Early Blight

Early blight is *Alternaria* blight is worldwide in distribution but only in important foliage disease of potatoes in few areas.

Symptoms

Brown, angular, necrotic spots marked internally by series of concentric rings from on leaves and to a lesser extent on stems. Leaf lesions are seldom circular because they are restricted by larger veins. They develop usually around blossom time and become increasingly numerous as plants mature. Lesions are first form on lower leaves. They may join, cause general yellowing, leaf drop or early death. Tuber rot is dark coloured, dry and leathery

Susceptible variety, usually early maturing, may show severe defoliation. Later maturing varieties may appear resistant. Plants under stresses that hasten maturity (such as adverse environment, warm, humid weather, other diseases, or nutritional deficiency become susceptible and die earlier than normal.

Control

Control is by providing conditions for vigorous growth throughout the season, especially irrigation and fertilizer side dressing. Organic fungicides sprayed on the foliage reduce spread of early blight.

Wart Disease

Wart or black wart is widely distributed in temperate and tropical regions with cold and rainy climate. In money cases considerable yield loss were observed where vigortis race of fungus occur and resistance plant has not been introduced.

The disease is often associated with powdery scab and does not develop in warm or in dry soils.

Causal Organism

Synchytrium endobioticum

Symptoms

Tumours growth of any size upto several centimetres may develop on stem stolen and tubers. Symptoms usually develop below ground but under wet conditions they may appear on stem and foliage. Initially tumours are white to brownish or of the same colour with abnormal tissue. Tumours blacken with age and may rot due to secondary organisms. Aboveground warts are green radish or purple, depending on variety.

Control

1. Varieties exist that are resistant to races of the fungus should be used.
2. Reduce incidence of wart with long crop rotations (Five year) in combination with resistant varieties.
3. Prohibiting the shipment of tubers, particularly seed from infected regions, is effective to limit disease spread.
4. Seeds of resistant varieties grown in infested soil can not spread the disease as well as movement of infested soil or manures adhering to tuber surfaces on farm machinery or other equipment.

Common Scab Disease

Scab is a common tuber defect in all potato growing regions of the world except where soils are very acid. The causal organism has been introduced into most potato growing area.

Symptoms

A variety of lesion develop on tubers. They may be superficial or reticular deep or pitted or protuberant. They are variable in size and shape and may join so that the entire surfaces become affected.

Control

1. Maintain high soil water levels compatible with good potato growth during tuber enlargement.
2. Avoid repeated crops of potatoes or there scab susceptible plants such as beet and fleshy-rooted crucifers,
3. Scab resistant varieties are useful.
4. Maintain soil levels at 5.0 to 5.2 with acid farming fertilizers or sulphur. Avoid heavy lime applications and preferable use of dolomatic, lime where needed.

Stem Canker or Black Scurf

The fungus is causing stem canker and black scurf is present nearly all soils because it has a wide host range, survives in plant debris and as sclerotia is easily disseminated on tubers. It will grow over a wide range of temperature. It causes considerable damage to emerging sprouts when condition do not favour rapid emergence, such as cold and wet soil.

Symptoms

Symptoms are varied. Lesions appear on sprout tuber tips cause a delay or failure to emerge. Slightly sunken brown cankers of irregular size and shape affect stolen and stems at or below the soil line. Cankers may girdle to stems and results in aerial tuber formation, plant wilt and death Girdled stolen may fail to produce tubers. Hard superficial dark brown or black sclerotia of irregular size and shape form on the tuber surface. A white mycelia mat may develop on the stem base but does little harm to the plant.

Control

1. Sclerotia are long lived in soil, only long rotations with cereals and grasses reduce disease incidence. Shallow planting of well sprouted tubers reduces exposure time of sprouts in the soil.
2. The disease can be reduced by application of soil fungicides such as PCNB (Pentachloronitrobenzen) mixed into the planting band of soil. Seed tuber treatment is effective for reducing seed borne inoculums and when soils are not heavily infested. It is not useful in heavily infested soils.

Potato Leaf Rolls Viruses (PLRV)

PLRV is most important potato virus and is common in all countries. Yield loss is up to 90 per cent.

Symptoms

Primary symptoms caused by current season infection are rolling of upper leaves especially of leaflet bases. These leaves tend to be upright and are generally pale yellow with many cultivars they are tinged purple, pink or red. Late infection may not cause symptoms and some cultivars develop necrosis of the vascular system. Secondary symptoms of plant grow from infected tubers of sub species tuberosum are rolling of basal leaves. Rolled leaves are stiff and leathery, sometimes tinged purple on undersides. Subspecies and genus reacts differently; marginal and interveinal chlorosis especially of upper leaves marked upright growth often severe stunting. Rolling of lower leaves is often completely lacking.

Natural transmission is by aphids in a persistent manner and through infected tubers.

Control

PLRV can be controlled by sclerotia of healthy plants and elimination of diseased plants through rouging in seed production. Systemic insecticide will decrease spread of aphids within the crop but will not prevent infection by viruliferous aphids from other field. It is only potato virus known to be eliminated from tubers by heat treatment.

Mosaics

Mosaic symptoms may be caused in leaves of potatoes by several different viruses singly or in combination. Some of these are potato virus X (PVX), Potato virus S (PVS) and potato virus M (PVM) as well as PVV and PVA. PVX causes yield loss above 10 per cent, the extant varying according to strain and potato cultivar, it is transmitted through infected tubers by contact (not by aphid), and normally causes mosaic. Infection may be mild in some cultivars and is frequently latent. Virulent strain may cause crinkling. Some cultivars are hypersensitive to certain strains, reacting with top necrosis.

PVS is common and may cause mild symptoms. It has little effect on yield. It is transmitted through infected tubers, by contact, and by aphids in some strains. Infection is normally latent although some cultivars react with mild mosaic or faint vein banding. A few sensitive cultivars react with severe browning, necrotic spotting or even leaf drop. PVM is less common and perpetuated through infected tuber and transmitted by contact and aphids. It is cause severe mosaic and leaf crinkling under certain environmental condition.

Control

It is controlled by clonal selection during seed multiplication. Rouging is useful only when obvious symptoms develop.

Black Leg and Soft Rot

Black leg of the potato plant and soft rot of the tuber are widely distributed disease, especially harmful in humid climates. *Erwinia carotobora* usually occurs in warm climates, *E.c.* spp. *atroseptica* occur in cool climates and *E.chrysenthemionly* in hot climtes.

Symptoms

Blackleg occur at any stage of development when moisture in excessive. Black slime lesions develop in moist weather often progress up on stem from the soft rotted mother tuber. New tuber some time rotted at the stolen end. Young plants are commonly stunted and erect. Yellowing and upward rolling of leaflets may occur, finally wilting and death of plants occurred.

Causal Organisms

Erwinia carotovora, E. atroseptica

Control

Avoid planting in wet soils and do not over irrigate. Harvest tuber when ripe, handling gently and do not exposed to sun. Tuber must be dry before storage or shipping; some varieties are more resistant than others.

Bacterial Wilt

Bacterial Wilt or brown rot is the most serious bacterial problem of potatoes in warm regions of the world. It often limits production.

Symptoms

Initial symptoms of wilt may affect one side of a leaf first, or one branch and not another. Later symptoms are severe wilt and death. A vascular strand darken and in cross section a gray to brownis yellow, exudes, except in mild cases which may confirmed by the flow of milky white strands from a stem, section immersed in clear water. A greyish bacterial slime may ooze out through the eyes or stolen end of tubers and from vascular ring of cut tubers. Either aerial or tuber symptoms may occur alone, but usually former is followed by the latter. Latent tuber infection occurs when infected seed is planted in collocations or in tubers infected late in the growing season.It develops rapidly at high temperatures.

Control

Control crop rotation is most effective with the potato race-3 but difficult with race-1 that also affects many other crops and weeds.

Diseased seed causes the most severe bacterial wilt infection and results in spread to non infested soil. Spread also occurs by water flow along rows fro, field to field. Root to root contact also transmits the bacterium.Soil survival varies considerably, usually 1 to 4 years for race-3. Often longer for race-1.

Chapter 11

Tomato (*Solanum lycopersicum*)

Early Blight of Tomato

This is common disease of tomato occurring all over India and other tomato growing countries. The disease is caused by the fungus *Alternaria solani* (Ell. and Mart.) Jones and Grout. The pathogen attacks the foliage causing characteristic leaf spots and blight

Symptoms

Small pale brown spots first appear on leaves. Lowest leaves are attacked first followed to upper parts of the plant. Concentric rings on leaves surrounded by chlorotic areas are the characteristic symptom of the disease. In dry weather spots become hard and leaves are curled but in humid weather big rotted patches are formed and leaves may fall down. Similar patches are developed on stem resulting in the whole plant to collapse. The rotting may reach upto the tuber. Affected plants develop less and smaller size tubers.

Pathogen

Fungus *Alternaria solani* (Ell. and Martin) Jones and Grout is the causal pathogen. The pathogen belongs to family Dematiaceae, order Hyphomycetales and sub-division Deuteromycotina. The conidiophores of the pathogen emerge through stomata from the spots and bear conidia. The conidia are beaked, muriform with 5-10 transverse septa along with a few longitudinal septa. Conidia germinate within 35-45 minutes at 28-30°C temperature. Other hosts of pathogen are tomato, chillies, *Atropha belladonna, Cymphomandra betacea, Hyascymus albus, H. niger* and *Nicotiana alata.*

Disease Cycle

The mycelium of the pathogen remains viable on dry infected leaves for 10-12 months. Conidia are also viable upto 17 months at room temperature. Initial infection therefore takes places either by mycelium or conidia surviving in soil on dead plant debris. Contamination of seed tubers with conidia or mycelium may be another source of primary inoculum. The infection starts from lower most leaves and further spread of the disease takes place from the conidia formed on these leaves. The conidia are disseminated by water, wind and insects. Under favourable conditions of rains followed by warm and dry weather the disease spreads very rapidly.

Control

Destruction of plant debris after harvest and use of certified seed are important measures to avoid infection. Regular sprays with fungicides at an interval of 10-21 days depending on severity of the disease should be done. Fungicide Dithane Z-78 @ 2lb/100 gallons of water gives best result. Other fungicides like Dithane M-45 (0.2 per cent), Blitox-50 (0.25 per cent), *etc.* have also been recommended for disease control. Resistant varieties like Kufri Navin (for hills), Kufri Sinduri and Kufri Jivan (for planes) should be used.

Late Blight of Tomato

The disease occurs on tomato in all tomato growing area. The disease is caused by *Phytophthora infestans* (Mont.) de Bary. The disease occurs in some parts of North India particularly in hills and also in Karnataka state. Some species of *Solanum* and *Lycopersicon* are the collateral host of the pathogen. Certain species and varieties of tomato are resistant to late blight. The disease is controlled by timely spraying of fungicides such as zineb, nabam, maneb *etc.*

Symptoms

The disease usually appear late in the season than early blight. Brownish black areas develop on leaves which can cover the whole leaf surface. The disease spreads very fast when high humidity is there and cover the whole leaf within 1-4 days. The disease can cover all the leaves and stem of the infected plant and sometimes the whole crop is blighted. The fruit can also be infected during harvesting or in storage. Most of the fruits infected plants are become rotted showing brown to purple discoloration on the skin of fruits. Moist atmosphere is suitable for disease.

Pathogen

The pathogen, *Phytophthora lycopersici* (Mont) de Bary is the causal pathogen of the disease. The pathogen belongs to family Pythiaceae, order Peronosporales and sub-division Mastigomycotina. The mycelium is endophytic and intercellular. Sporangiophores develop from the mycelium and come out through stomata or lenticels of tubers. Sporangia are pear shaped and papillate and develop at the tips of sporangiophores. The sporangia germinate and form secondary sporangia which produce zoospores upon germination. The zoospores are biflagellate and cause infection to the host. Sexual reproduction is not known. Several physiologic races are known in the pathogen.

Disease Cycle

Primary infection most likely comes through seeds particularly under Indian conditions. The mycelium from contaminated seeds grows upward in the stem and cause sporulations on small dwarfed shoots. Infection of foliage takes place from the spores produced in primary lesions. The sporangia are spread through wind, water and leaf eating insects to the healthy crops in the region. The spores are washed out to the soil from affected leaves and infect fruits. The fruits also got contaminated when come in contact of infected leaves during harvesting or in contact with infected tubers in the storage.

Control: Use of certified seeds, sanitation, delayed harvesting, storage in cool, dry and well aerated stores are the major precautions to avoid infection. Spraying of crops with fungicides starting well ahead of disease appearance time can provide good control. Dithiocabamates (Zineb, Maneb *etc.*.) are more widely used to control the disease. Spray with Dithane Z-78 and Dithane M-45 @ 2-2.5 kg/1000 liter of water/ha have been recommended. Early sprays of Bordeaux mixture (4:4:50) and late spays at an interval of 15-21 days with Bordeaux mixture (6:6:50) should be given.

Wilt of Tomato

The disease is found throughout the world and is serious where tomato crop is grown in the same field continuously.

Symptoms

Clearing of the leaf veins and chlorosis are the initial symptoms of the disease. Petioles and leaves of affected plants droop and wilt. In severe cases black discoloration of vascular tissues can be seen if the roots and stem are split open.

Pathogen

Fusarium oxysporum f. lycopersici (Sacc.) Snyder and Hansen is the causal pathogen of the disease.The pathogen belongs to family Tuberculariaceae, order Moniliales and sub-division Deuteromycotina. The pathogen hyphae are found in the vascular tissues and are inter and intracellular. The pathogen produces both macro and microconidia.

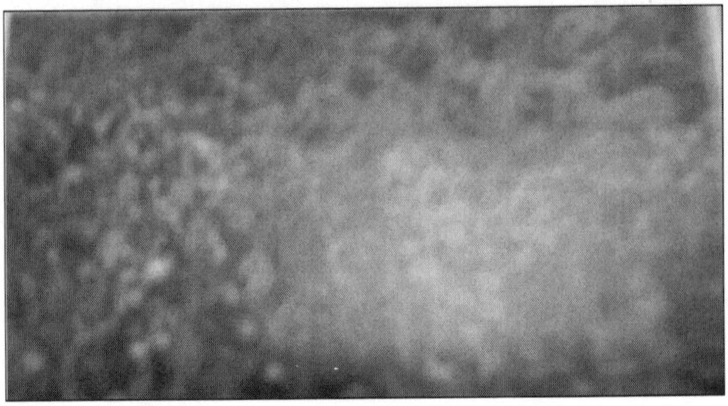

Disease Cycle

The pathogen remains saprophytically in soil and attack the root region when crop is planted in field or plants are raised in nurseries. After the infection the pathogen moves upwards and multiply rapidly in vascular bundles causing interference with upward movement of the nutrients which causes wilting of plants. Probably factors like plugging of vascular tissues, action of phytotoxins and pectic and cellulotic enzymes acting together cause the wilting of plants.

Control

Since the pathogen is soil borne, control is difficult. The best control of the disease is to grow exotic varieties such as Rutgers, Kanora and Roma *etc.* which have been reported to have resistance.

Root Knot Disease of Tomato

Root knot disease is found in most of the countries on various vegetable crops such as potato, tomato, brinjal, chilli, okra, bhendi, cucurbit *etc.* The losses caused by this disease are from 20-75 per cent in tomato and 17-81 per cent in brinjal.

Symptoms

The affected tomato and brinjal plants show unthrifty development and stunted growth if the infection occurs during early stages of plant growth. Leaves become yellowish green to yellow and sometimes show scorching along the margins. The roots show galls of various sizes and shapes. Presence of root galls is the most characteristic symptom of the disease.

Pathogen

The disease is caused by the nematode pathogens of the genus *Meloidogyne*. The common species are *M. incognita, M. javanica, M. arenaria* which commonly attack tomato and other vegetables. The male nematodes are vermiform and 1.2 to 1.5 mm x 30-36μm in size. The females are pear shaped and measure 0.15 – 0.25 mm wide at the base. Upto 600 eggs are laid by each female.

Disease Cycle

A large number of females are found in the root galls which survive in the soil on plant debris. The second stage larvae are liberated in the soil. Sandy light soil is best for their movement. These larvae can be killed at higher temperature of 40-50°C and excess of soil moisture. The larvae are attracted by root exudates and thus come in contact of the roots. They gathered in a mass around roots and cause infection by penetrating roots of healthy plants. The male and female nematodes develop depending on the number of larvae per unit area of the host tissues. If they are more crowded mostly male develops but if only few larvae are there than females will develop. The reproduction is mostly parthenogenetic. A temperature of 25-28°C is best for infection and gall formation by nematodes.

Control

Cultural practices such as crop rotation, fallow soil, sanitation, deep ploughing during summer, soil solarization and flooding of soil are important to avoid infection by the nematodes. The organic soil amendment such as addition of neem cake, karanje cake @ 25 quintals/ha reduced the root knot incidence. The use of soil fumigation with nematocides is the best method of control but these chemicals are highly toxic and expensive. Resistant varieties such as SL-120, Nimetex, Y-220, PAU-1(7-3-1-4), Hissar Lalit *etc.* for tomato and Vijay, black beauty, T-3, S-149 *etc.* for brinjal should be used. Transgenic plants and biological control are promising and may be exploited for nematode control in future.

Damping Off of Tomato

Damping off is a common disease of nurseries of tomato. The disease is found worldwide.

Symptoms

The young seedlings are killed before they emerge or the seeds failed to germinate. Young seedlings can be attacked at any time during germination. The infection spreads very fast killing the invaded tissues resulting death of seedlings. This is known as 'pre-emergence damping off. Next is the 'post emergence phase' in which seedling can be killed at any time of their growth till the stem becomes hard enough to resist invasion. Usually roots or stem at or below ground level are attacked. Infected tissues become water soaked. The infection spreads and basal part of the stem become thinner than rest of the stem fall down on the ground. The infection in seedlings continues till they are completely dead. The disease usually radiates from the point of infection to nearby plants and dead seedlings can be seen in large areas of nursery beds.

Pathogen

The damping off is caused by several fungal pathogens but most common are *Pythium* spp.such as *P. debaryanum* Hesse, *P. aphanidermatum* (Eds.) Fitz., *P. ultimum* Trow and *P. arrhenomanes* Drechsler. Several othe fungi and bacteria can also cause similar symptoms such *Phytophthora, Sclerotinia, Fusarium, Rhizoctonia etc.*

Pythium produces white mycelium on which sporangia are developed. The sporangia either germinate directly by germ tube or by producing a balloon like secondary sporangium known as vesicle. In this vesicle over 100 zoospores are formed and they swim after their release and then rounded off to form a cyst. The zoospores create new infection after germination. The sexual stage develops as oogonia and antheridia. The oogonium after fertilization develops thick wall and known as oospore. The oospores can survive in adverse conditions and this is the resting stage. The oospores germinate in the same way as sporangia.

Disease Cycle

The pathogen is soil borne. Spore germinates by germ tube and directly penetrates into the seed or young seedlings when comes in contact. Proteolytic enzymes break the protoplast and in some cases the cell wall is disintegrated by cellulolytic enzymes causing death of seeds or young seedlings. The disease is more serious in poorly aerated and poorly drained soil. As the infection progresses the sporangia begin to appear, followed by sexual spores inside or outside of host tissues. The movement of infection is very fast and can quickly rot even the fleshy vegetables or fruits in the field storage, in transit and even in market when healthy and infected fruits come in contact.

Control

Sterilization of soil by steam or dry heat will control damping off in nurseries. Use of seed protectants such as blitox-50 (soaking seeds in 0.20 per cent solution) for tobacco seeds, Agrosan-GN or agrosan (seed dressing @ 0.3 per cent by seed weight) for tomato seeds and mercuric chloride (1: 1000 solution dipping for 5 min.) for cucumber seeds is important for disease control. Soil treatment of nurseries by formalin (1 : 50) drenching or spraying several days before nursery planting is important. Nursery beds should be raised with light soil having good proportion of sand with high nitrogen dose. The seeds should be sown thin and only light irrigation at frequent intervals shoud be given. Spraying of seedlings with fungicides like metalaxil, captan *etc.* would be required if soil is heavily infested with the fungus and was wet for long time.

Virus Diseases of Tomato

Tomato plants are susceptible by large number of virus diseases such as leaf curl, black ring, bushy stunt, golden mosaic, mild mottle, mosaics, yellow leaf curl *etc.* Tomato spotted wilt, mosaic, leaf curl and a cucumovirus are important for India.

Tomato Yellow Leaf Curl

Symptoms

The affected tomato plants remain stunted with chlorotic and puckered leaves. Curling of leaves with bright yellow colour is the characteristic symptoms of the disease.

Pathogen

The causal virus is a geminivirus and belongs to family geminiviridae. The virons are geminate, 20 nm in diameter and 30 nm in length. The genome is single stranded DNA, circular and distributed in both the particles. Virions are found in phloem. The virus is transmitted by an insect vector, *Bemisia tabaci,* a whitefly in a persistent manner. The virus is also transmitted mechanically and by grafting. Several plant species in family solanaceae, leguminaceae and malvaceae are susceptible to virus infection.

Disease Cycle

The disease spreads through white fly vector when the crops are overlapping or from symptomless plants like *Malva parviflora.*

Control

Control of white fly vector by safe chemicals or to grow crop in vector-free period are the only preventive measures at present.

Tomato Indian Leaf Curl

The disease is present in tomato throughout India and is a major threat to tomato production.

Symptoms

Affected plants remain stunted, with small chlorotic puckered leaves. Affected plants look bushy in the field and hardly bear any fruit. The mild or severe symptoms of the disease depend on the variety and strain of the causal virus.

Pathogen

The causal virus is a geminivirus and belongs to family geminiviridae. The virus has large number of hosts in several plant families. The virions are geminate, 20 x 30 nm in size. The particles are found in phloem parenchyma and in nuclei. The genome of the virus is single stranded DNA, circular and in two parts known as DNA-A and DNA-B. The virus is transmitted by white fly, *Bemisia tabaci* in a persistent manner.

Disease Cycle

Both virus and its vector have very wide host range. The inoculum is present throughout the year either on cultivated crops or on weeds. The virus is not seed

borne. The inoculum already present on other hosts is brought to the new crop by insect vector

Control

Control of vector by insecticides, trap plants and yellow mulches and use of less susceptible varieties are the only management practices at present.

Tomato Spotted Wilt

Tomato spotted wilt virus has been reported from most of the countries infecting one crop or another. The virus has been reported on potato, sunflower, groundnut, tobacco *etc.* In India, the virus is known to infect sunflower, groundnut and several plants of family leguminaceae. However, there is no authentic report on tomato.

Symptoms

Necrotic or chlorotic local lesions, systemic wilting, necrotic spotting, streaking, mosaic, mottling, malformation of leaves, vein yellowing, ring spots, line patterns and colour breaking of flowers are the variety of symptoms found on various hosts.

Pathogen

The causal virus is tomato spotted wilt virus which belongs to Tospovirus group in family Bunyaviridae. The virus is transmitted by several species of *Thrips and Frankliniella. T. tabaci* is the major vector and the virus is transmitted in a persistent manner. The virus multiplies in vector and has a transovarial transmission. It is also transmitted mechanically and by grafting.

The virus particles are isometric, enveloped, 85 nm in diameter. Virions contain 5 per cent nucleic acid, 70 per cent protein and 20 per cent lipid and 5 per cent carbohydrate. The genome of the virus is ssRNA. The virions are distributed in all parts of infected plants.

Disease Cycle

The inoculum survives throughout the year on one host or the other from where it reaches the healthy plants by thrip vector. The inoculums are also present in thrips and they can infect the plant without fresh acquisition on infected plants.

Control

Control of insect vectors and development of virus resistant transgenic are the main methods to control tomato spotted wilt virus.

Tomato Mosaic

The disease has a very wide host range and is found throughout the world.

Symptoms

Systemic leaf mosaics with leaf narrowing are the main symptoms of the mosaic disease.

Pathogen

The causal virus is tomato mosaic virus which belongs to tobamovirus group. The virus particle is rod shaped, usually straight with a modal length of 300 nm and width 18 nm. Virions contain 5 per cent nucleic acid and 95 per cent protein. The genome is unipartite and is ssRNA. The virus does not have a vector but is transmitted by mechanical inoculations, grafting, and by contact of diseased and healthy plants. It is transmitted externally seed infection.

Disease Cycle

The primary inoculums come via seed and then spread by contact. Infection normally occurs during transplanting.

Control

Seed treatment with sodium hypo chloride or any other chemical to remove external inoculums from seed and use of resistant cultivars are the only measures to manage the disease.

Cucumovirus Disease

Symptoms

Severe leaf malformation, leaf mottling, leaf reduction sometimes showing shoe string symptoms, dwarfing and malformed fruits are the main symptoms of the virus on tomato.

Pathogen

The cucumovirus is a member of family Bromoviridae and has a very wide host range. Virions are isometric, 29nm in diameter. The genome is ssRNA divided in several parts. There is non-genomic nucleic acid in the virions known as satellite RNA. The virus is transmitted by several species of aphids including *Myzus persicae* in a non-persistent manner. It is also transmitted by mechanical inoculations and in seeds of a weed, *Stellaria media.*

Disease Cycle

The virus spreads in healthy crop by the aphid vectors from the inoculums present either on crop plants or in weed host.

Control

1. Control of aphid vectors.
2. Use the virus resistant transgenic plants to manage the disease
3. Use seedling from healthy fields.

Chapter 12

Onion (*Allium cepa*) and Garlic (*Allium sativum*)

Purple Blotch of Onion

The disease is common and wide spread. It attacks both the bulb as well as the seed crop. Complete failure of seed crop was noticed at Dhaulakuuan (HP) in 1966-67 and at Bangalore (Karnataka). Losses of about 20-60 per cent in bulb yield have been reported from Punjab, Haryana and Maharashtra. The loss is depends on time of infection and stage of crop growth.

Symptoms

A small water soaked area developed on leaf and seed stalk, and turn brown. This spot enlarges and become zonate and purplish in colour. More or less leaves and flower stalk break at the point of infection. The margin is shade of purple or red, surrounded by yellow hole that extend upward and downward for some distance. Large lesion develops which form leaf shrivelling and dies. Bulb and seed development is adversely affected. Diseased umbels produce shrivelled seeds which lack viability. At harvest time affected bulbs of the plant shows decay.

Causal Organism

Alternaria porri (Ellis), cif

Conidiophores arise single or in group, they are strait or flexuous, often geniculate, septate, pale or mid brown. Conidia are muriform solitary, straight, curved, obclavate, ellipsoid, tapering to a beak, mid golden brown colour 8-12 transverse septa and zero to several longitudinal septa. Each cell of conidium is capable to germinate by germ tube. Chlamydospores are also reported in this genus.

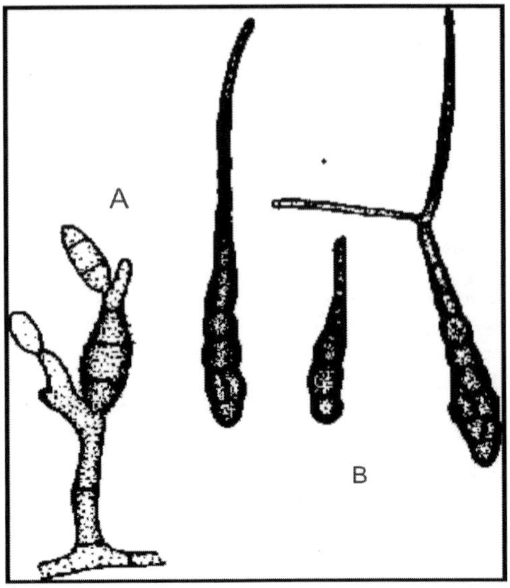

Environmental Condition

Moist weather, Optimum temperature 25°C, Relative humidity 100 per cent for few hours.

Survival

Crop debris, infected seed, diseased bulb and infection on other host plant such as garlic.

Dispersal

Soil, water, air borne conidia and thrips.

Control

1. Crop rotation
2. Cleaned seed
3. Field should be well drained
4. Seed treatment through thiram@ 2.5g/kg seed
5. Spray mancozeb (0.25 per cent) or Difolatan (0.3 per cent) alongwith Malathion (2.50g a.i./ha)
6. Ziram and nebam at 10 days intervals after first lesion appearance.
7. Resistant variety- Red Creole, Udaipur 101, Kalwan Red.

Stemphyllium Blight of Onion and Garlic

The disease is very common in northern and eastern part of India on onion and garlic leaf as well as tuber and stalk. Disease intensity ranges between 20- 90 per cent in seed crop and 5-50 per cent in bulb crop.

Symptoms: Infection usually starts on radial leaves after transplantation at 3-4 leaf stages. The infection starts during December or early January in north India. It attains maximum intensity in the blossoming period. Flecks soon develop into elongated spindle shaped to ovate elongate diffusive spots surrounded by a characteristics pinkish margin. Initially yellow to pale orange streaks develops on middle of leaves or flower stalk are on one side. At later stage, they develop in both sides as circular lesion to spindle shape to ovate spots. Later it turns brown in colour. In advance stage lesion may girdle and kill leaf and seed stem. Concentric zones may develop within lesions. Sometime during high humidity spore may develop on lesions. The lesion of spot coalescing and blighted the leaves, but rarely affects the bulbs. The pathogens survive normally on dead and drying lesion of the leaf spots. The spot are also surrounded by pinkish margins.

Environmental Condition

The disease is favoured by heavy dew, foggy and rainy weather. Optimum temperature for disease development is 21°C.

Causal Organisms

Stemphyllium vesicerium (Wallr) Simson.

Perithecia (Sexual stage) appear in blighted area in small, black pin head like raised bodies towards the close of season. Conidiophores dark, mostly simple with darken terminal swelling, short to long bearing single terminal conidia or successive conidia on new growing tips. Conidiophores are proliferating through old conidial scar. Conidia (Porospore) are dark with cross longitudinal septa variable in shape globose to broadly ellipsoid or ovoid smooth verrucose to echinulate, air borne and blastic. The spores are air borne.

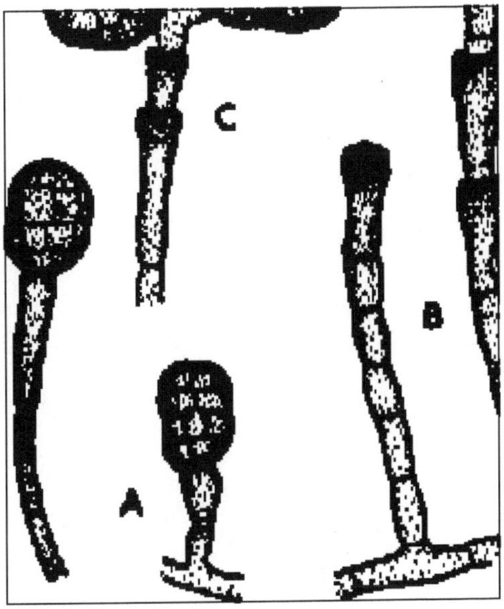

Control

1. Field sanitation, collect and burnt crop residue which minimize disease infection.
2. Spraying of mancozeb @ 0.25 per cent along with Sandovit @0,5ml/lit.
3. Resistant/tolerant line of onion IC-321 should grow.

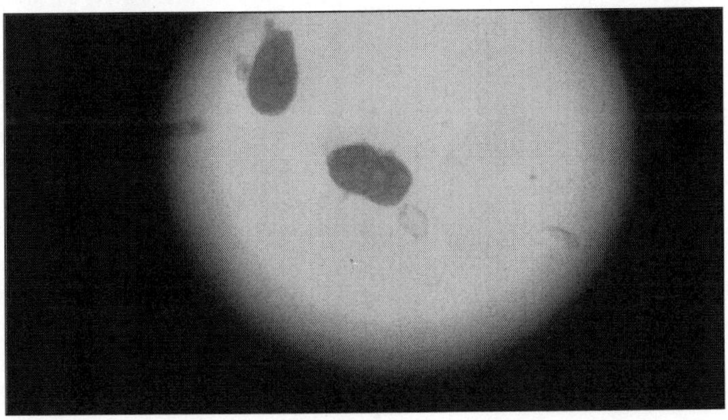

Downy Mildew of Onion and Garlic

Downy mildew was first reported on *Allium* species from Kashmir valley in an epidemic form by Berkeley in 1841 with yield losses between 60-100 per cent. The disease is severing in relatively cool and moist weather. It was first reported in India from Kashmir valley in an epidemic form during 1974-76 with yield losses between 60-100 per cent and subsequently reported from Haryana and Himanchal Pradesh. The disease is responsible for sever yield losses in Kangra valley.

Symptoms

The disease appeared first on leaves either by perennial plant of onion or through infected bulbs. Violet tinted downy growth over the entire surface of leaves appeared in humid weather, these later turn greenish grey as the disease advances. The infected lesions are oval to cylindrical in shape, variable in size and often paler than the normal yellow green of the leaf and stem. When the leaves are infected between the centre and the tip, the leaf droops at the point of the lesion and the tips dies. Foliage and flower stalks wither, bulbs become moist and spongy. The severity of the disease depends on the environment. Plants are seldom killed, but bulb growth is reduced and the bulb tissue is inclined to be spongy and of poor keeping quality.

Causal Organism

Peronospora destructor (Berk) Casp

The sporangiophores are non septate, variable in length from 122 to 820µ, and swollen the base to a diameter of 7 to 18µ. Branching is monopodial with sub acute to acute sterigmata bearing pyriform to fusiform sporangia. The sporangia are thin walled, sub hyaline and slightly appellate at the distal end. They do not form zoospores but germinate by germ tubes. The oogonia formed intercellular, are 43 to 54µ in diameter and germinate by germ tubes.

Survival

The oospores of pathogen overwinter as systemic mycelium in bulbs and infected plants in mild winter area. It is also survive in the form of oospores in soil. Oospores can survive on the surface of infected seeds.

Dispersal

Seed, seedling, soil and air current.

Environmental Condition

Sporangia are produced in high humidity over range of temperature 4 to 25°C and optimum is 13°C. Sporangia are developing during night and maturing early in the morning, which disseminated during the day. Alternation of darkness and light influences the sporulations.

Control

1. Inorganic and organic fungicides shows reduction in the amount of disease.
2. Mixture of Dithane Z-78 and sulphur reduces the disease development.
3. Ridomil MZ or Difolatan (0.2 per cent)and Aliette give good control.

Smut

Causal Organism

Urocystis cepulae Frost

This is one of the most important diseases which is difficult to control and reported all major onion growing area in India. It is mostly confined in Karnataka state at present

Symptoms

Infection appears on young leaves as dark, slightly thickened lesions and forms readily visible streaks within the tissue of older leaves, leaf sheath and bulbs. Black powdery spore mass (teliospores) is released from the streaks which continue to develop throughout the life of plant. Some bending and twisting of leaves is noticed. Sometimes the infection goes to the base of plant and most of the plant dies within 3-4 weeks.

Survival

Pathogen can survive in soil for many years through spore balls. If host become available spore germination take place otherwise they can remain dormant for a long period. Mycelium can also survive in plant debris.

Dispersal

Pathogen dispersed through bulb, soil, seed, Plant debris and, seedling.

Control

1. Seed treatment with Thiram or Capton (2g/Kg seed)or organomercurials and furrow applicationof these fungicides in seed beds reduces seedling infection
2. Nursery soil should be treated with fumigants to raise healthy seedling
3. Application of formaldehyde in the soil before planting
4. Use resistant varieties

Storage Rot

Onion is stored at the farm level by the growers in traditional stores without proper curing. The storage arrangement is inadequate and unscientific resulting 15-60 per cent spoilage. Various pathogens associated with rotting bulbs include: *Sclerotium rolfsii,Aspergillus niger, Alternaria solani, A.cepicola, fusarium solani, Fusarium campttocera, cephalosporium curtipes, Penicellium paxilli* and number of bacteria. Incidences are mostly favoured by lack of proper aeration, high humidity, lack of proper grading of bulbs and storage of immature bulb.

Management

1. Ammonia releasing compounds *viz.* Ammonium carbonate and bicarbonate, ammonium sulphate, diphenylamine or other sulphur dioxide producing fumigants is effective for minimizing storage losses.
2. Dipping of bulbs for ten minutes before storage in capton (0.2 per cent), thiram (0.2 per cent) or aureofungin (100ppm) or acetic acid reduce storage loss.

3. Grow recommended varieties suited for different season and location. Lorge size bulb do not stored because they become more rottage.

Other Diseases

Neck Rot

Botrytis allii Munn: Leaf blight is more common, Infection take place through injured tissues as small white speck on leaf. Infection and its further development take place between 10-24°C temperature. It is favoured by high humidity.

Control

Spray of capton, ferbam, or ziram protect the foliage against infection.

Black Mold

Aspergillus niger Tiegh

Smudge

Clletotrichum circinans (Berk) Vogl.

Black Mould Disease of Onions

The disease is more prevalent in tropical regions of country causing post-harvest blemishes and rotting of onion grown and stored in hot climate. The disease is common in storage.

Causal Organism

Major pathogen is *Aspergillus niger* and others include *Aspergillus fumigates* and *A. alliaceus.*

Symptoms

Fungus infects neck of bulb and rarely penetrates from sides and basal end of bulb giving post- harvest blemish appearance leading to rotting in presence of other organisms like bacteria and Fusarium sp. *etc.* As a result of growth of fungus, white mycelium produced on the infection site results in to brown to sooty black mass of conidia. These spore mass can easily be rubbed off. Scales are shrivelled and brittle in appearance. The disease is favoured by humid conditions and is checked under dry and cool environment.

Management

1. Care should be taken in proper curling, drying, sorting of diseased bulbs and grading to avoid injury before safe storage.
2. Bulbs may be dusted with calcium carbonate to reduced rotting before storage.
3. Bulbs may stored in containers fumigates with nitrogen trichloide at the rate of 430 mg/m^3 during transit.
4. Seed treatment with either benomyl + thiram @2.5 + 2.5 g a.i./Kg seed or in hot water (15 minutes immersion at 60°C temperature).

Onion Smudge

This disease is of historical importance because of the development of principles of biochemical constituents in disease resistance in this pathosystem. The disease was reported in England in 1851 for the first time and since then its occurrence has been noticed throughout Europe, USA, Japan, Argentina and India. Smudge is present in onion, shallots, lacks and *Allium fistulosum* not on garlic. It is a serious disease of white onions but red onions resistant to it.

Symptoms

The disease may appear in the field, during the transit and also in storage. The disease is characterized on the outer scales of the bulbs as minute spots consisting of stromata just beneath the cuticle or on neck and green leaves that are first dark green in colour and later form black dots. Stroma may be scattered over the surface of the lesions, but they frequently are formed in circular, concentric rings. Normally the outer scales are affected but inner scales, the lesions are small, sunken and yellow. In humid weather, Pinkish mass of spores develop on the black stroma.

Causal Organism

Colletotrichum circinans (Berk.) Voglino. The fungus belongs to division Eumycota, subdivision Deuteromycotina, Class Coelomyctes, order Melanconials and family Melanconiaccae.

The mycelium of the fungus is septate, branched, hyaline when young and become darker and thicker with age. The thick walled hyphae intertwine and form dark green to black stromata on which acervuli are formed. Acervuli contain short, hyaline conidiophores along with dark coloured setae which are 1 to 3 septate and 80-315 µm long. Conidia (3-6 x 14-30 µm) are fusiform. Nonseptate, hyaline to slightly ochraccous, somewhat curved and obtuse at the apex. Conidia germinate by producing one or two germ tubes which form appresoria after coming in contact with hard surface.

Disease Cycle

The fungus overwinters on the onion bulbs, sets and in the soil. The spores produce mycelium which first penetrates the outer scales. It is only under the most favourable conditions that the organisms enters actively growing tissue. In 5 to 6 days, the mycelium becomes established to form acrevuli filled with spores. The fungus is disseminated by wind, splashing water, and though tools and clothing. The pathogen is transported long distance on the onion bulbs and onion sets, and possible on the seed.

Garlic

Leaf Blight of Garlic

The disease has been reported from Kulu valley (H.P.) by Singh and Sharma (1977).

Symptoms

It was more severe in fields irrigated with sewage water. Infected spots are small, yellowish, circular to oblong initially later become spindle shaped and dirty white or grey in colour. Attacks garlic leaves producing lesionsThey break in the middle and finally wither at premature. Infection does not extend to scales.

Pathogen

Stemphylium botryosum Wal,Perithecia (Sexual stage) appear in blighted area in small, black pin head like raised bodies towards the close of season. Conidiophores dark, mostly simple with darken terminal swelling, short to long bearing single terminal conidia or successive conidia on new growing tips.

Control

1. Field sanitation, collect and burnt crop residue which minimize disease infection.
2. Spraying of mancozeb @ 0.25 per cent along with Sandovit @0,5ml/lit.
3. Resistant/tolerant line of onion IC-321 should grow.

Rust

The disease appeared in a severe form in south west Punjab during 1987—88. Both uredo and teleutospores were noticed on garlic whereas only uredosori were noticed on onion. Incubation period was 10 days. Other spore stages were not found.

Storage Organs

Discolored or dead areas that penetrate deep into storage organs such as tubers, bulbs, corms, *etc.* is caused by many fungi and some bacteria. Dry rots often are caused by fungi, which also may produce mycelia or spores. Soft rots usually are associated with bacteria such as *Erwinia* spp. and can be accompanied by strong, repulsive odors. Often, bacterial soft rots follow rots caused by fungi, making diagnosis difficult. Other internal problems (for example, ring rot of potato) can be caused by bacteria or by several viruses.

Scurfy, dead surface tissue may be caused by a variety of fungus like organisms (for example, powdery scab of potato). Galling of storage organs can be caused by fungi or nematodes.

Symptoms and Management of Important Vegetable Diseases

Sl.No.	Name of disease	Causal organisms	Chief symptoms	Management
1.	Club root of crucifers	*Plasmodiophora brassicae*	Enlarged roots appearing spindles or clubs of both the main and lateral roots. The aerial parts of plants usually wilt	1. Use well drained diseased free plots for planning 2. Crop rotation avoiding any type of cruciferrous crop 3. Apply PCNB formulations as Terraclor, Quintozone and Brassico @30Kg/ha one week before planting in soil
2.	Damping off Tomato, brinjal, crucifers and cucurbits	*Pythium* sp. *Phytophthora* sp. And *Rhizoctonia* sp.	The rapid death and collapse of very young seedling in the nursery and field	1. Treat the seed with Captan @ 1:500, drench seedlings in the nursery beds with Captan @1:500 in water.
3.	Root rots of cucumber, crucifers, brinjal, chilli etc.	*Pythium, Rhizoctonia, Phytophthora* and *Botrytis*	Disintegration or decay of parts or all of root system of plant, whereas the tissue become pulpy carmine red and corresponding above ground parts turn yellow and die.	1. Quintozene (PCNB) or Thiophanate methyl can be applied locally to plant base 2. Diseased plant debris must be destroyed
4.	Soft rot or Dry rot of Cucurbits, green bean, potato, cabbage and chilli	*Pythium* sp. *Phytophthora* sp. and bacteria *Erwinia*. sp	Maceration and disintegration of fruits, roots, bulbs, tubers and fleshy leaves. It is consists of watery or slimy decay accompanied by an offensive odor.	1. Avoid planting in wet soil 2. Do not over irrigate 3. Resistant varieties should be grown
5.	Late blight of potato and tomato	*Phytophthora infestans*	Extremely rapid browning of leaves, branches and floral organs resulting in their death, under humid weather condition water soaked spots develop on the leaves and enlarge to become indefinite in shape.	1. Spray the crop with Dithane Z-78 or Indofil M-45 Plus Zinc sulphate @2.5 g/litre water 2. Spray Ridomil MZ in early stage at above dose
6.	Downy mildew of cucurbits pea onion, garlic	*Pseudoperonospora, Pernospora pisi*	Chlorotic or necrotic area develop on leaves when plants are 3-4 leaf stage, In humid condition fungus produces a violet downy mildew over entire surface of leaves, in dry condition white spots appear on systematically infected leaves.	1. Seed treatment with Apron @ 6.0g/kg seed 2. Spraying of Ridomil @0.25 per cent 3. Spray of copper fungicides of Blitox-50 @0.25 per cent

Contd...

Sl.No.	*Name of disease*	*Causal organisms*	*Chief symptoms*	*Management*
7.	Rhizopus soft rot of vegetables	*Rhizopus* nigricans	Infected organs appear water soaked with soft finally dryup and disintegrated	1. Use disease free tubers for planting 2. Avoid excessive moisture
8.	Leaf blotch of turmeric	*Taphrina maculans*	Small oval, rectangular or irregular brown spots on both sides of leaves but numerous on upper sides of leaves. They are arranged in rows along the veins.	1. Use 3-4 spray of Dithane Z-78 @ 0.2-0.35 per cent or Dithane M-45or Blitox or Bavistin from disease initiation to maturity 2. Use disease resistant cultivars like China, Jawali and NDH-98
9.	Powdery mildew of cucurbits, cabbage and pea	*Erysiphae cichracearum* and *E.polygoni*	Grayish white floury powdery mass of the fungus appear on the young leaf, branch and pods of plants	1. Most effective fungicide is sulphur dust @25-28 Kg/ha 2. Spray with Thiovit @0.2 per cent solution
10.	Leaf spot or black rot of cucurbits	*Cladosporium cucumerinum*	Symptoms appear as grayish spot with brown periphery, sometimes angular with yellow halo	1. Seed treatment with Thiram or captan @ 2.5g/kg seed 2. Spray the crop with Indofil M-45 or Dithane Z-78 @ 2.5 g/litre water three times at 15 days intervals
11.	Early blight of potato and tomato	*Alternaria solani*	Symptoms appear first as spots on the leaflets. These are dark brown to black and leathery appearing necrotic tissue. Concentric ridges often form to produce a characteristic symptoms	1. Provide conditions for vigorous growth through out the season 2. Spray Indofil M-45 @ 0.2 per cent or Blitox-50 or Difolatan @ 0.25 per cent at 15 days intervals.
12.	Cercospora leaf spot on eggplant, pea and tomato	*Cercospora* sp.	Roughly circular leaf spots with radish- purple border formed and later their center become ashy-grey	1. Spray the crop with copper fungicide e.g. Fytolon 2-3lb per 1000gallons of water per acre.
13.	Anthracnose of chilli and bean	*Colletotrichum capsici* and *C. gloeasporium*	A necrotic and sunken ulser like lesion formed on the stem, leaf, fruit and flower	1. Seed treatment with Ceresan @ 0.2 per cent of the weight of seed followed by Ziram spray @ 0.2 per cent has been found most effective against the disease

Contd...

Sl.No.	Name of disease	Causal organisms	Chief symptoms	Management
14.	Fruit rot or die back of brinjal and tomato	*Colletotrichum phomoides*	Extensive necrosis of twigs beginning at their tips and advance their base. Fruits which are ripe or being to ripe, small, water soaked, circular spots shows as slight depressed at any point on fruit, they are very numerous and closely crowed.	1. Rotation of seedbed 2. Zerlate is most effective fungicide 4 spray are required after 30days of planting at 10-12 days intervals with addition of any adhesive for better coverage of the fruit
15.	Onion smudge	*Colletotrichum circinans*	The spots are roughly circular and occasionally contain stomata arranged in concentric rings. Under moist conditions the stomata bear acervuli, which contain prominent black setae readily distinguished with a lens.	1. The artificial drying of bulbs immediately after harvest checks the disease under dry conditions of storage 2. Protection of harvested crop from rains, rapid drying, thorough curing and storage under well ventilated conditions 3. Red onions are generally resistant
16.	Phomopsis blight of brinjal	*Phomopsis vexans*	Stem is girdled slightly above the soil line, the plant topples and dies. Brown, round or oval spots develop on the leaves, becoming irregular in shape as they enlarge up to 3 cm diameter. Lesion contain pycnidia and develop dark brown zone. The fruit is attacked become the decay and being soft and watery and later black, pycnidia develop aboundantly.	1. In nursery spray Ziram or Captan @ 0.2 per cent at 7 days intervals and in transplanted field spray Maneb or Zineb @0.25 per cent at 10 days interval
17.	Wilt of tomato	*Fusarium* sp. and *Verticillium* sp.	Shoot lose their turgidity and droop because of a distortions of vascular system of root	1. Use 10-15g *Trichoderma viride* in per Kg FYM before transplanting. 2. Use disease resistant varieties of tomato
18	Black leg of crucifers	*Phoma lengum*	Spots appear on the leaf as inconspicuous, indefinite, pallid areas that gradually become well defined spot with ashen-gray centre, where pycnidia are scattered irregularly. On the stem purplish spot appear, Root system gradually destroyed and new root emerge above the lesion which keep plant alive	1. The reliable seed treatment is with hot water at 50°C temperature for 30 minutes 2. Phoma free seed should be used 3. Three year crop rotation fallow

Contd...

Sl.No.	Name of disease	Causal organisms	Chief symptoms	Management
19.	Sclerotinia disease of vegetables	*Sclerotinia sclerotiorum*	White fluffy mycelia growth on leaves, stem, flower and fruit, which soon enlarge and compact with resting bodies or sclerotia.	1. Drench with 1-2000mercuric chloride at the base of plant is helpful 2. Avoid short crop rotation with crucifers
20.	Rust of pea	*Uromyces pisi*	Many small lesions on leaves, stem and some time on pods usually of a rusty colour. It produces uredia and telia on pea and some other legume	1. Three spray of IndofilM-45 @ 3 Kg/ha at 10 day interval or one spray of Baycor @500g/ha
21.	Onion smut	*Urocystis cepulae*	The lesion consists of dark, slightly thicked area in leaf and cotyledons in several millimeters. The lesion formed in plants at all stages often rift to expose black powdery masses of spores.	1. Drenching of a mixture of lime and sulphur in the furrow with the seed, or apply a solution of formaldehyde in 10 gallons of water applied at the rate of 1 gallons to a 150 foot furrow, for nurseries dress the seed with 1 lb of Agrosan per 10lb of seed.
22.	Purple blotch of onion and garlic	*Alternaria porri*	Small, white sunken lesion with purple centers rapidly enlarge and eventually girdle leaf or seed stem after some time superficial masses of fungus appear on the lesions	1. Seed treatment with Thiram @2.5g/kg seed 2. Spray fungicides like Indofil M-45, Dithane Z-78, Foltaf @0.25 per cent at 15 days intervals. Sticker Savdovit @0.1 per cent may also be added with fungicidal solution.
23.	Angular leaf spot of cucumber	*Pseudomonas lachrymans*	The disease appears on leaves, stems and fruits as small water soaked spot, spot enlarge about 3 mm in diameter, becoming tan on the upper surface and gummy or shiny on the lower surface and assuming an angular shape as the lesion is delimited by veins	1. Use disinfected seeds (Treated with sodium chlorite or chlorinated lime) treat with copper fungicide or Mancozeb from propagation onwards wetting plant well 2. Spraying of mixture of Copper fungicide and Mancozeb @0.25 per cent
24.	Bacterial leaf spot of tomato and pepper	*Xanthomonas vesicatoria*	Small, dark greasy spots appear on leaflets and stems of seedlings and when numerous, they cause severe damage, Most damaging phase is infection of immature fruits, Small, water soaked spots appear on exposed surface of the green fruits	1. Rotation of seedbed and transplanting field at least 3 years 2. Use Bordeaux mixture for spray 3. Resistant variety should be sown

Contd...

Sl.No.	Name of disease	Causal organisms	Chief symptoms	Management
25.	Bacterial wilts of cucurbits	*Erwinia tracheiphila*	Bacterial wilts affect plants by causing sudden wilting of foliage and veins and finally death of plants. After infection, the bacteria invade and progress within the tracheal vessels a viscid matrix in the vessels plugging to xylem	1. Pathogens are transmitted by beetle hence management of beetle should be adopt. 2. The best combination has been found Zerlate plus rotenone, applied as a dust
26.	Bacterial wilt of tomato	*Clavibacter michiganese*	Spotting or wilting of leaflet at the outer and lower part of tomato. Wilting and leaf curl occurred upwards and downwards but not fall off	1. The disease can be reduced by application of Brassicol (PCNB) in soil @ 20-30 Kg/ha 2. Seed treatment with Agallol or Aretan @ 0.25 per cent for 5-10 minutes
27.	Black rot or black vein of crucifers	*Xanthomonas campestris*	V-shaped chlorotic blotches on the margins of leaf, that progress towards the midrib of the leaves	1. Use disinfected seeds (Treated with sodium chlorite or chlorinated lime) treat with copper fungicide or Mancozeb from propagation onwards wetting plant well 2. Spraying of mixture of Copper fungicide and Mancozeb @0.25 per cent
28.	Bacterial soft rot of vegetables	*Erwinia carotovora*	Small water soaked lesion appear on vegetable fruits and enlarges rapidly in diameter and in depth and become soft and mushy. Disease caused severe damage in storage and transit. Black rot invasion occurs through leaf scars and bacteria cause a soft rot of the fleshy pith.	1. Sanitation must be practiced, walls of store should be cleaned and disinfected with formaldehyde or copper sulphate @ 2.5 percent solution, Leafy vegetables may be pre cooled to 4-6°C
29.	Comman scab of potato and sweet potato	*Streptomyces scabies* and *S. ipomaeae*	Symptoms appear on tuber, first they consist small brownish and slightly raised spots and later enlarge coalesce and become corky.	
30.	Tomato mosaic virus	*Tomato mosaic virus*	Infected plant consist various degree of chlorosis, curling mottling dwarfing, distortion and blistering of leaves.	

Contd...

Sl.No.	Name of disease	Causal organisms	Chief symptoms	Management
31	Potato leaf roll	*Potato leaf roll virus*	Upward Rolling of the margins progress toward the midrib of leaf. The tissue assumes rigid, leathery texture and this gradually extends to the whole top. Plants become dwarfed.	
32.	Cucumber mosaic	*Cucumovirus*	Young leaves become mottled, distorted, wrinkling and edges become curl downward	1. Affected plants should be removed from the field 2. Spray of insecticides should be done to check the aphids
33.	Yellow vein mosaic of Okra and bean	*YVMV*	The first visible symptoms of the disease appear as yellow spots and scattered on the lamina surface. They are mostly round in shape. In yellow mottle, the spots are diffused and expend rapidly. The pots are deformed and shriveled.	1. Use resistant varieties like Pusa A-4, Vijay, Parmani. Kranti, Pusa Sawani and Hissar Unnat.
34.	Root Knot Nematodes	*Meloidogyne sp.*	Plant shows stunted growth, leaves are yellowish green to yellow, which tend to drooping and then there is sudden wilting. The root bears spherical to elongated galls	
35.	Little leaf of Brinjal	*Mycoplasma*	Small leaf with bushy growth. The affected plants do not bear fruit.	1. Rouge out the diseased plants 2. Antibiotic treatments with tetracyclines have been found effective to some extent. 3. Control of insecticides are the only recommendations against the disease.

Diseases of Spice Crops

Chapter 13

Ginger (*Zingiber officinale*)

Soft Rot of Ginger

Soft rot of ginger is common occurrence in all ginger growing tracts in India. Butler (1907) first recorded this serious disease of ginger in Gujarat. The disease appear in epidemic form, the losses is caused very enormous usually 10-15 per cent reduction in yield where crop was grown in low land and 5 to 6 per cent in upland. In wet year loss was estimated more than 80 per cent.

Symptoms

Symptoms are produced both on above ground and below ground part of plant. In above ground, drying of the leaves starting from the margin can be seen in the field. Water soaked spots in collar region are produced by the fungus. In sever condition whole plant get collapsed. The disease manifested itself on leaves,

which turn pale green in comparison to healthy ones. Yellowing notice on the tip of the leaves and gradually turn yellow and the yellowing spread down wards of the leaf blade and leaf sheath more along the margin than at the centre. The yellowing is followed by withering and drying and finally dead. The dead leaves ultimately droop off and hang down along the stem till the entire shoot becomes dry. Below ground fungal infection decays the rhizome as well as roots. Poorly developed roots and less formation of the rhizome are the other symptoms. In severe cases the basal portion of the plant becomes watery and softtand the rhizome, which is discoloured gradually decomposes, farming water masses enclosed by the touch ring of rhizomes, such plant easily pulled out from the rhizome.

Causal Organisms

Pythium aphanidermatum, P. myriotylum, P. vexans

The hyphae of the fungus are colourless, branched forming irregular swelling, non septate hyphae on branched mycelium and form long tapering zoosporangium by apical swelling, when there is enough moisture in the environment zoosporangium is not cutoff from the rest of the hyphae only protoplasm and nucleolus transferred in to the vesicle. Which protoplasm goes into the cleavage and formed small segment, each of the segments formed into zoospore, which escape from the vesicle by rupture of the cell wall? Zoospores are kidney shaped, slightly depressed and measures 8-12µm x 6-8µm. Zoospore swim some time and encysted to formed walled structure then take rest and germinate by direct germ tube. Sexual reproduction takes place within the tissues of affected plants. Oogonia are formed lateral or intercalary it is spherical shape, thin walled hyaline. Antheridia may be formed terminal or intercalary and arise branched different which bear oogonia. Oogonia and antheridia fuse and formed oospores, oospore germinate by taking the rest.

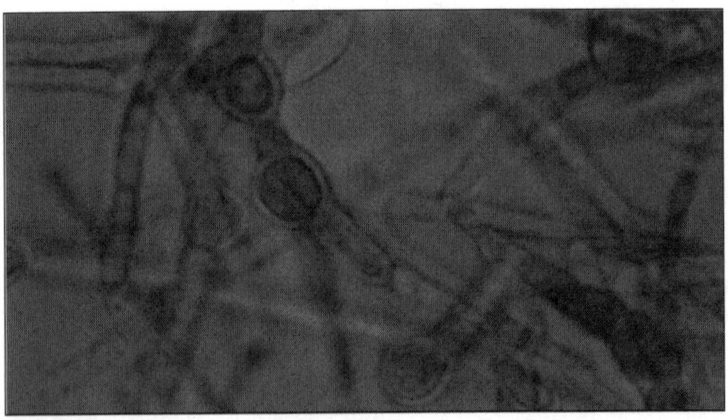

Survival and Spread

Two ways they are carry over:

1. Through disease rhizome
2. Through resting structure of causal organism, such as chlamydospore

3. Pathogen survive in soil, they have saprophytic ability to live in soil for long time.
4. Mimegralla flies also play role in transmission of the pathogen.

Environment

High soil moisture is highly conducive to the disease development. The fungus multiplies with buildup of soil moisture with the onset of south west monsoon. Younger sprouts are the most susceptible to the pathogen.

Control

1. Use healthy and disease free rhizome for sowing. Seed multiplied by tissue culture is very suitable for sowing.
2. Grow the crop in light soil and avoid stagnation.
3. Dipping of rhizome in 0.25 per cent cereson for 30.0 minutes.
4. Rhizome treatment with echlomezole (Citridiazole) and soil drenching
5. Soil drenching with Blitox and fytolan @ 0.04 per cent or Mancozeb @0.3 per cent
6. Maintain good drainage and avoid low lying field
7. Intercropping with okra gives satisfactory yield
8. Resistant variety- Rio-de-Janiero, Maran, Vengura and Gorubathan
9. Bio-fumigation with mustard and cabbage also gave satisfactory results
10. Application of *Trichoderma harzianum* along with neem cake @ 1.0Kg/bed helps in preventing the disease

Yellow Disease or Wilt of Ginger

Yellow or wilt disease is an important and destructive disease of ginger in Queensland, Hawai and some part of India.

Symptoms

The first symptom of this disease are conspicuous yellow of margins of lower leaves which gradually spreads over the entire lamina and later shows typical

wilt and ultimately dries off but leaf do not fall on ground. The basal part of the affected plant becomes soft and watery. The shoot can be easily pulled out from the affected mother rhizomes. Disease rhizomes show creamy discoloration of the various regions. Development of rhizome badly affected. The patches of disease appeared in the field.

Causal Organism

Fusarium oxysporium f.sp. zingiberi

Mycelium delicates, white or peach coloured with purple tinge, sparse to abundant micro conidia borne on simple phialides cylindrical, straight to curve. Macro conidia are borned on more branched conidiophores or on the surface of sporodochia. They are thin walled 3-5 septate fusoid, pointed at both ends.

Survival and Spread

Disease perpetuates through infected rhizomes, which act as primary inoculums in new field. Secondary spread through mechanical means like water, soil survives as chlamydospore. Temperature favours 15-38 °C, optimum temperature 23-29 °C, Relative humidity 87-95 per cent.

Control

1. Soaking of seed rhizome in Dithane M-45 (0.3 per cent) or Benlate or Bavistin 0.25 per cent for two hours with soil drenching.
2. Crop rotation with Jower
3. Seed treatment with *Trichoderma viride* @ 4.0g/kg rhizome.
4. Use *Bacillus subtilis* of biological control

Bacterial Wilt

The disease was first reported from Mauritius. In India, it is first reported from Madras (1941) now it is found all ginger growing area.

Symptoms

The first conspicuous symptoms are mild drooping and curling of leaf margins of the lower leaves which spread upwards. Yellowing starts from the lowermost leaves and gradually progresses to the upper leaves It appears as water soaked, linear streaks or patches on the collar region of pseudo stems followed by yellowing or bronzing of leaf margin starting from lowermost leaf. Progress gradually upward and downwards later the leaves become flaccid with intense yellowish- brawn margin and droop showing typically wilt symptoms. Affected pseudo stem and rhizome slime with disintegration and give offensive smell, when small pieces of infected rhizome is put to water, it shows turbid bacterial ooze.

Causal Organism

Pseudomonas solanacearum or Ralstonia solanacearum

Bacteria are non spore forming, no capsulated, gram negative, short rod shaped, motile with single polar flagellum. Colonies on solid medium are small, irregular, round white in reflected light.

Survival and Spread

Pathogens survive through infected rhizomes and weeds. Soil and seed borne disease that occurs during south west monsoon.

Control

1. Crop rotation
2. Use healthy rhizomes
3. Rhizome treatment with 0.6 per cent mercurial chloride
4. Soil fumigation with methyl bromide (1.3kg/10m^2)
5. The seed rhizomes may be treated with Streptocycline@ 200ppm for 30 minutes and shade dried before planting.
6. Once the disease is notices in the field all beds should be drenched with Bordeaux mixture @0.1 per cent or copper oxychloride @0.2 per cent.

Leaf Spot

1. *Colletotrichum zingiberi*
2. *Helminthosporium maydis*
3. *Phyllosticta zingiberi*

Leaf spot is mostly caused by *Phyllosticta zingiberi* and disease is notices on leaves from July to October. The disease starts as a water soaked spot and later turns

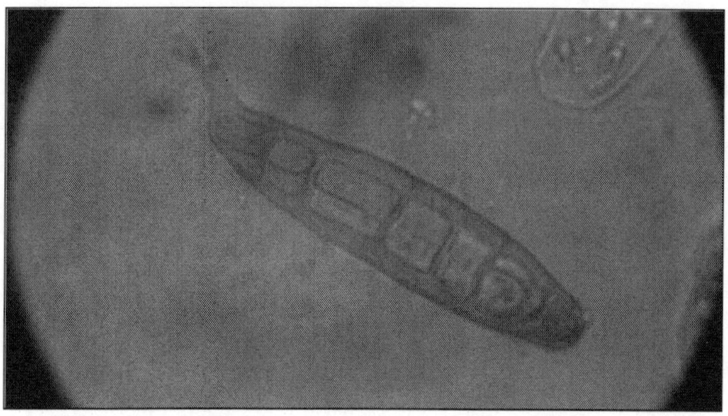

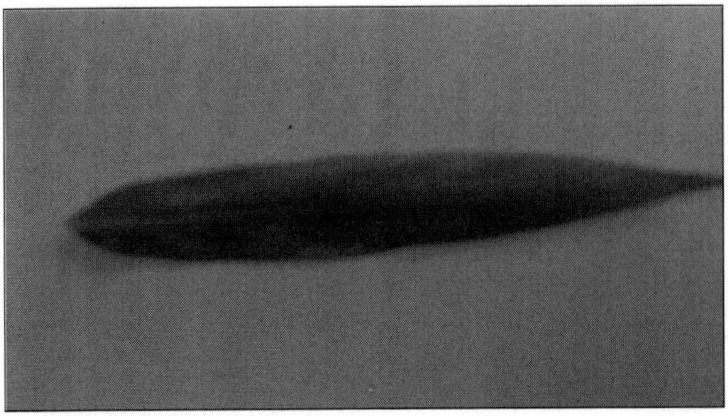

as a white spot surrounded by dark brown margins and yellow halo. The lesion enlarges and adjacent lesions coalesce to form necrotic areas. The disease spreads through rain splashes during intermittent showers. The incidence of the disease is sever in ginger grown under exposed conditions.

Control

The disease can be controlled by spraying Bordeaux mixture @ 1 per cent or Mancozeb @ 0.2 per cent.

Nematodes

Root knot- *Meloidogyne* spp.,Burrowing- *Radopholus si*mili, Lesion- *Pratylenchus* spp. are important nematodes pests of ginger. Stunting, chlorosis, poor tillering and necrosis of leaves are the common aerial symptoms. Root galls and lesions that lead to rotting are generally seen in roots. Infested rhizomes have brown, water soaked areas in the roots and in the outer tissues.

Control

1. Treating infested rhizomes with hot water at 50°C for 10 minutes.
2. Use nematodes free seeds like IISR Mahima variety.

Chapter 14

Turmeric (*Cucurma longa*)

Taphrina Leaf Spot

Taphrina leaf spot of turmeric is commonly prevalent in India, wherever crop is grown. This disease is often found to cause much damage to the crop in southern India. The crop is widely cultivated in the northern states as well and found infected with the Taphrina leaf spot disease. It was first described by Butler (1911) as occurring in Gujarat, Saharanpur (Uttar Pradesh) and Rangapur (Bangladesh).

Symptoms

The large number of disease spots appears on both surfaces of the leaves, but spots are more numerous on the upper surface. The spots first appear as small, oval, rectangular or irregular pale yellow discolouration soon become dirty yellow and deepened to old gold colour. The spots are 1-2mm and coalesce freely. In severe

cases the plant present a scorched appearance.The infected leaves are not distorted but they have reddish brown to yellow and remain alive despite of heavy infection, which resulted high yield loss.

Causal Organism

Taphrina maculans Butler

It is supposed to be an obligate parasite. But applying special technique with Pavgi and Upadhya (1964) successfully cultured the pathogen on PDA media. The hyphae of the parasite are intercellular, embedded in the cuticle and cell walls of the epidermis. They send branched or lobbed shaped haustoria into the cell of epidermis and hypodermis. The moss of hyphae accumulates in the entire epidermis of spots. Hyphae mass pass downward and split the epidermis and spread out in the inner wall. The development of asci take –lace on the both surface of the leaves and all the outer cells become asci and they mature in group. The asci are saclike or lovate or round or flattened shape measuring 20-30µm. Normally eight ascospores are formed in each ascus. The ascus are hyaline unicellular ovoid and 4-6µm x 2.25µm in diameter.

Survival and Spread

The fungus survives as ascogenous cells on leaf debris in the soil among fallen leaves and rhizomes of turmeric as formed ascospores or blastospore. The disease severity depends on inoculum potential available in the soil. Primary infection occurs on the lower leaf in October and November at 80 per cent relative humidity and 21-23°C temperature. Secondary infection occurred by ascospores discharge from asci under cool humid conditions.

Control

1. Disease can be controlled by spray of Bordeaux mixture or Aureofungin 2.5mg/ml water or Zineb @0.1 per cent or Dithaine Z-78 @0.20-0.35 per cent.
2. The leaf blotch can be controlled by Hexaconazole @ 0.20 per cent with three spray
3. Use resistant variety China,Jawali and NDH-98
4. Long crop rotation
5. Deep summer ploughing

Colletotrichum Leaf Spot

The disease was first reported from Coimbatore district of Tamil Nadu by Mc Rae (1913) as pathogen *Vermicularia curcmae* but latter Ramakrishnan (1947) come to conclusion that this disease was caused by *Colletotrichum capsici* because *C. curcuma* is the same species on the basis of conidia and cross inoculation tests. It is prevalent all over India but maximum in Tamil Nadu, Andhra Pradesh. Disease reduced the yield loss by 62.6 per cent but oil and curcumin may slightly increased.

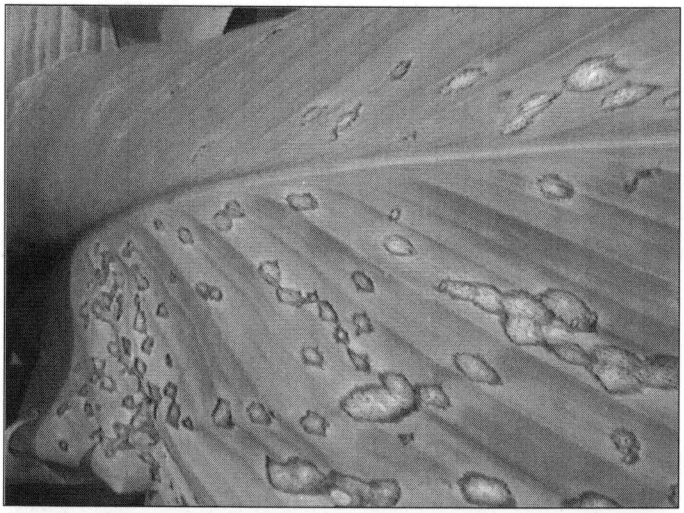

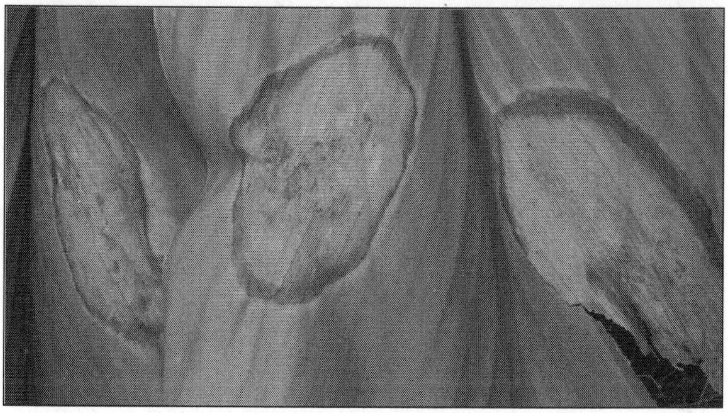

Symptoms

The disease usually occurs in August and September when high humidity prevail in the atmosphere. The pathogen attacks on the leaves which are usually confined on leaf blades and may occasionally be extended to leaf sheaths. Initially they appear elliptic to oblong diffuse chlorotic spots on the leaf in variable size on both surface but more on upper surface. The spots gradually increase upto 4-5 x 2-3 cm. The fully develop spots is grayish- white at the centre with a brown margin all around a yellowish ill defined halo around the infected area. The production of yellow halo is responsible for the production of toxin (endotoxin). Two or more spot coalesce to form necrotic patches. Sever infection most of leaves dry up and caused yield loss more than 50 per cent. When crop is mixed with pigeon pea the infection is severe because it provides shady, moist environment and spread to most of the plants in the field. The rhizome does not develop well.

Causal Organism

Colletotrichum capsici

Mycelium are both inter and intracellular septate hyaline at first and later become pale brownish. They accumulate in epidermal cells which increase in pseudo paranchymatous cell which increase in size, rupture cuticle and grow cushion shaped acervuli on the surface, setae and conidiophores develop. The setae are septate, tapering, long radiated scattered all over the acervulus. The conidia are borne single on the top of conidiophores. They are crescent shape, one celled, hyaline measuring 17-31 x 3-4µm. Conidia remain in the base around the setae and surrounded by mucilaginous substance which hold all together on drying. Conidial mass is pinkish. It is germinating by producing germ tube in one or both end of conidia and infection taking place. They also produce chlamydospore which survives in unfavorable environmental conditions.

Survival and Spread

The conidia spread long distance with air and survive on the rhizome as dormant stage. Conidia disseminate in the wet weather and infection take place in high humid environment.

Control

1. Two spray of Bordeaux mixture @ 1.0 per cent at the time of disease appearance.
2. Vitavax, Plantvax and Auriofungin are also effective against Colletotrichum leaf spot.
3. Captan and Dithane Z-78 can be used for management of this disease by spray at 30 days intervals.
4. Use tolerant variety or resistant variety such as TS-2, TS-4, TS-79 *etc.*

Leaf Blight

The disease was first observed by Maiti *et al.* (1980) during monsoon months of 1977 from north-eastern states of India. The disease was found infected in large area of the field in every year.

Symptoms

Disease was seen as water-soaked spots varying in shape and size on the lower leaves which increase in size during warm and humid weather, resulting in blighting of entire leaf. The blighted leaf area is divided into various sectors which are characteristic symptoms of the disease. Fungal growth is visible on the undersurface of the leaves.

Causal Organism

Rhizoctonia solani Kuhn.

The fungus hyphae are initially hyaline but turn light brown. They produce white globose to sub-globose, oval or cushion like sclerotia develop on the medium which turns deep brown. Sclerotia formation starts on medium after 4-6 days at 28±1 °C measuring 0.5-3mm. But it survives from 15-30 °C temperature.

Survival and spread

Fungus is a soil borne pathogen. Initial infection starts from soil Secondary infection taking place through rain water.

Humid and moist weather are congenial for infection.

Control

Soil drenching of Pentachloronitrobenzine (Brassicol) is found to effective for management of disease.

Rhizome Rot in Dtorage

Rhizome rot is causing by *Aspergillus flavus, A.niger, Cladosporium cladosporoides, Sclerotium rolfsii.*

Nematodes

Root knot- *Meloidogyne* spp., Burrowing- *Radopholus simili*, are two important nematodes pests of turmeric. Stunting, chlorosis, poor tillering and necrosis of leaves are the common aerial symptoms. Root galls and lesions that lead to rotting are generally seen in roots. Infested rhizomes have brown, water soaked areas in the roots and in the outer tissues.

Control

1. Treating infested rhizomes with hot water (50 °C) for 10 minutes.
2. Use nematodes free seeds.
3. Increased organic content of the soil.

Chapter 15

Fennel (*Foeniculum vulgare*)

Powdery Mildew

This disease is prevalent in Rajasthan and reported in all fennels growing area in India. Infection of the disease can be seen in the month of February and March and it causes heavy yield loss.

Symptoms

White powdery patches of mycelium developed on the upper surface of leaves, stem and other parts of plants. The high temperature with high relative humidity are favorable to disease development.

Causal Organism

Leveillula taurica (Lev.) Am.

Erysiphe polygoni. D.C.

Erysiphe umbelliferarum f. sp. anethi

The mycelium is both intercellular and superficial hyphae at first grow in the intercellular spaces and ramify in between the mesophyl cells of the leaves and from houstoria to the entire cell. Conidiophores arise from the internal mycelium and emerge through stomata either single are in groups. They are thin walled septate branched bearing single conidium at its tip. Conidia are single celled, cylindrical, irregular shaped, hyaline. Conidia germinate by forming germ tube and attached with host by appressoria. Cleistothecia appear on these superficial mycelium at later stage as small black specks immersed in the mycelia mats having numerous appendages, which are colourless or brown, each cleistothecium have 10-40 asci,which are large cylindrical to ovate and long stalk. Fennel (*Foeniculum vulgare* Mill.), belonging to the family Apiaceae, is widely cultivated and used as a culinary spice. In October 2014, fennel exhibited symptoms of powdery mildew with 100 per cent disease

incidence in a plastic greenhouse in Namwon City, Korea. Symptoms first appeared as white colonies, which exhibited abundant mycelial growth on the leaves, often covering the whole surface. Infected plants were unmarketable mainly owing to signs of white fungal growths on leaves and stems. Severe infections often caused leaf withering and premature senescence. A voucher specimen was deposited in the Korea University Herbarium (KUS-F28393). Hyphae were septate, branched, and 4 to 7 μm wide. Appresoria were multilobed to moderately lobed, and solitary or in opposite pairs. Conidiophores were cylindrical, 80 to 140 × 8 to 10 μm, and composed of 3 to 4 cells. Foot-cells of conidiophores were straight to sub straight, cylindrical, and 25 to 50 μm long. Singly produced conidia were oblong-elliptical to oblong, occasionally ovate, 30 to 50 × 12 to 16 μm with a length/width ratio of 2.4 to 3.6, had angular/rectangular wrinkling of outer walls, and lacked distinct fibrous in bodies. Germ tubes were produced on the perihilar position of conidia. Primary conidia were apically conical, basally truncate, and generally smaller than the secondary conidia. Cleistothecia were not found. These structures are typical of the powdery mildew *Pseudo-odium*, anamorph of the genus *Erysiphe*. The specific measurements and morphological characteristics were consistent with those of *E. heraclei* DC (Braun and Cook 2012). To confirm the identification, the

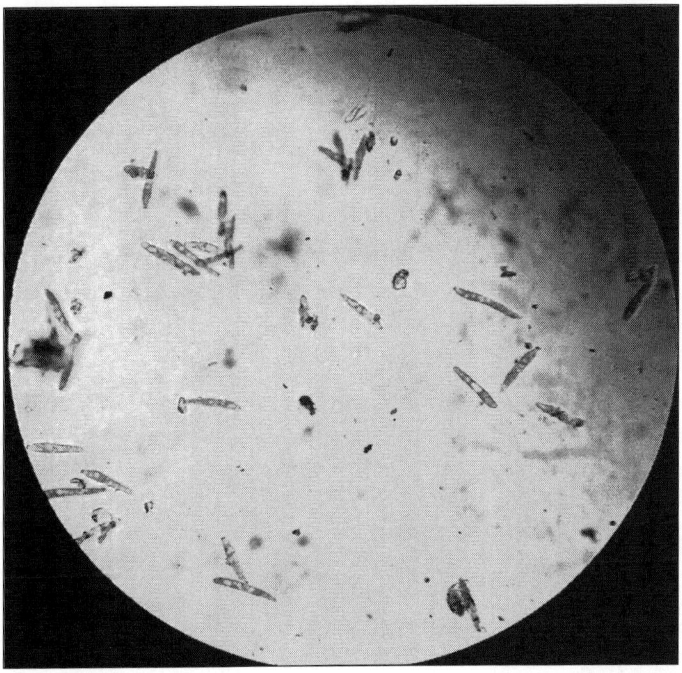

complete internal transcribed spacer (ITS) region of rDNA from KUS-F28393 was amplified with primers ITS1/ITS4 and sequenced. The resulting 700-bp sequence was deposited in Gene Bank (Accession No. KP729443). The obtained ITS sequence shared >99 per cent similarity with those of *E. heraclei* from apiaceous hosts, *e.g., Daucus carota* from China (KC480605) and *Anthriscus cerefolium* from Korea (KF111807). Pathogenicity was confirmed through inoculation by gently dusting conidia on the leaves of three healthy potted plants. Three non-inoculated plants served as controls. Plants were maintained in a greenhouse (26 ± 4°C and 70 ± 10 per cent RH) with a 12-h photoperiod. Inoculated plants developed symptoms after 5 days, whereas the control plants remained symptomless. The fungus present on the inoculated plants was identical in morphology to those observed in the field. Fennel powdery mildew caused by *E. heraclei* has been reported in many European countries, Egypt, India, Pakistan, Argentina, and Brazil (Rosa *et al.*, 2008; Farr and Rossman 2015). However, this is the first report of fennel powdery mildew caused by *E. heraclei* in Korea. Fennel is mostly grown using organic farming methods without chemical control options to produce fresh leaves and bulbs. Occurrence of fennel powdery mildew is a serious threat to quality and marketability.

Survival and Spread

Pathogen are obligate parasite which perennates through cleistothecia present on debris lost cropping season, which produce ascospores and infect the old leaves near the soil. Secondary infection occurs through the conidia and dispersed by air.

Control

1. Thiovit, sultaf and karathane are effective when applied twice by spraying with 0.1 per cent at 20 days interval.
2. Dusting of sulphar @28 Kg/ha is recommended for control.
3. Use resistant varieties
4. Summer ploughing is also recommended
5. Wettable sulphur @ 0.1 per cent can spray for management of disease
6. Early maturing variety should be grow for escape the disease.

Leaf Spot

Fennel has been reported to be infected by *Cercospora foeniculii* in Bihar, Kashmir, Rajasthan and other part of India. In India for the first time in 1929 Sydow and Mc Rae observed *Cercospora foeniculii* affecting leaves, petioles and stem of fennel. The disease is widespread and caused yield loss in Rajasthan. It is considerably reduce the foliage and poor development of fruit.

Symptoms

Brown to black lesion with pale centers is formed on the leaves stem, peduncles and seed. Severely infected plant shows blighted appearance and the yield is highly reduced. In case the crop infected in early stage, the yield loss enhances greatly even sometime no seed formation has observed. Light brown conidiophores are formed in clusters which emerge through stomata.

Causal Organism

Cercospora foeniculii

It is also accommodated in *Ramularia foeniculi*, pathogen invades only chlorenchymatous tissue in aerial parts. It grows intercellular in the palisade region by stomatal penetration.

Conidiophores emerge in fascicles are simple, subhyaline at the tip, darker below 30-40 um x 3.5-4.5 um. The conidia are hyaline 3-5 septa.

Survival and Spread

Pathogen is internally seed borne. Soil borne infected plant debris is a recurrence of the disease in next season crop.

Environment

High temperature and humidity favour the disease development.

Control

Spraying of colloidal suspension of sulphur @ 2.5 g/liter of water is recommended during February and March.

Ramularia Blight

Ramularia blight of fennel is most destructive disease of India

Symptoms

All above ground parts of plants are infected. Minute angular, brown to black spots are formed which later enlarge and get covered with grayesh white growth. The lower leaf get defoliated. The seed formed on diseased plants are shriveled and becomes black. This seed go for poor germination. Blight caused by *Ramularia foeniculi* was observed severe in Rajasthan. It is caused yield loss when occurred at flowering stage. The elliptical spot produced at lower side of leaf. In severe conditions, whole plants are look like with dusting of ash and caused heavy yield loss. It is only sporadic disease.

Environment

Conidial germination is favour at 12-22°C and 32°C temperature. Favourable conidition for disease development is sufficient water deposition on host during February and March. Wet and humid conditions predisposed to conidial germination and light helped for infection.

Control

1. Stop the irrigation for a week
2. Two spray of Dithane M-45 or Dithane Z-78 @0.8 to 1.0Kg per ha at the interval of 10-15 days
3. Difoletan and Mancozeb two spray at 10 days intervals @ 0.2 per cent was found most effective.
4. Benomyl @0.2 per cent or Tridemorph @0.1 per cent was found most effective against the disease.
5. Use resistant variety Ajmer fennel-1

Phyllody

Phyllody of fennel reported from India by Bhat *et al.* (2008). The disease caused by Phytoplasm which belong to peanut witches brown group.This disease is prevalent in Gujarat. The sugary liquid comes out in the umbels which make them sticky.

Control

1. Vector control by spraying with 0.2 per cent phosphomidon or Dimethoate @0.33 per cent or Quinolphos @0.05 per cent at 15 days intervals
2. Stop application of fertilizers and irrigation.

Chapter 16

Fenugreek (*Trigonell foenum-graecum*)

Powdery Mildew

Powdery mildew has been reported several part of India, it appear usually late in the growing season (February and March), therefore caused yield loss damage is less. If infection occurred early in February, resulted heavy yield loss. It has been reported to infect fenugreek from France, Germany, India, Iran, Italy and America

Symptoms

Erysiphae polygoni is characterized by the appearance of white floury patches on the both surface of leaves and other above ground parts of plant. Initially green foliage, tender stem and branches get infected. This white powdery mass contains conidia in large number. All the aerial parts exhibited the symptoms. Number of pods and number of seeds per plant get reduced. Localization of various metabolites in different places in diseased leaf and stems are observed. In severe condition plants get dried

Causal Organism

Erysiphae polygoni (DC)

Leveillula taurica (Liv.) var. *languinosa*

The mycelium of the fungus is entirely superficial only sending houstoria into the epidermal cells. The mycelium is branched, septate and forming a tangled mass. Conidiophore arises vertically on upper surface of the mycelia growth. They are erect, stout measuring 106.5-166.5μm in length. Each conidiophore bear the conidia in chain, mature conidia fall off from the tip towards the base. Conidia are hyaline barrel shaped, elliptical, and unicellular in varying size. Conidia germinate by germ tube in water. Optimum temperature for germination of conidia is 20-22°C upto 30°C. The development of powdery mildew favored to warm humid environment. *Leveillula taurica* favoured by 60-70 per cent relative humidity at 15-25°C temperature, it is severe in India Israil, Sudan,and Pakistan

Survival and Spread

Pathogen has survived in seed and soil. Ascospores formed in cleistothecia on the leaf surface towards the crop maturity as small black bodies immersed in mycelium mass that survive in plant debris of the crop. Secondary infection occurred by air through conidia.

Control

1. Dusting of sulphar @ 28 kg/ha at the time of flowering and second at the time of seed formation @ 13.5 kg/ha.
2. Perenox can used for disease management
3. Karathane and cosan @ 0.1 per cent is good for the management of this disease.
4. Hexaconazole @0.05 and Tridimefen @ 0.2 per cent may be used for the control of disease.
5. Deep summer ploughing
6. Sanitation of field and plant debris.
7. Closan @0.2 per cent or cosan or thiovit or sultab are found effective in controlling the disease of both pathogens by spraying twice at 15 days interval.
8. Resistant line GC-39,UM-32,GC-7,GC-20 and UM-34
9. Tolerant varieties are Rmt-305, Hissar Suvarna, Hissar Madhuri, Rmt-1 and Lam selection-1

Cercospora Leaf Spot

Cercospora leaf spot has been reported from Israil (Ming and Solely, 1959), North America (Leppik, 1960), Hungary (Varos and Nagy, 1972) and Canada (Zimmer, 1984). It is severe constraint for the production of fenugreek in India

Symptoms

The disease starts as very small reddish brown spots which spread on the leaves, stems young pods and seedlings and damaging the plants before they ripen. The spots are white at the center and brown at the periphery. Pathogen produces symptoms which become large spots on leaves, stem which caused severe defoliation and affects stem and pods. The spots are elongated on stem and petioles. The affected leaves drop off prematurely. Under favurable conditions the foliage appears fully blighted. Later the plant gets dried. Spots are amphigenous, round to semicircular about one centimeter in diameter, brown on the margined and white at the centre. The occurrence of disease adversely affects the market quality of green leaves.

Causal Organism

Cercospora traversiana Sacc.

Conidiophores, which emerge in fascicles are simple, sub-hyaline at the tip, darker below, 30-40µm x 3.5-4.5µm and have a number of denticulations, the conidia are acrogenous, acicula, attenuated at the apex, hyaline, 3-5 septate, measuring 80-125µm x 3.0µm.

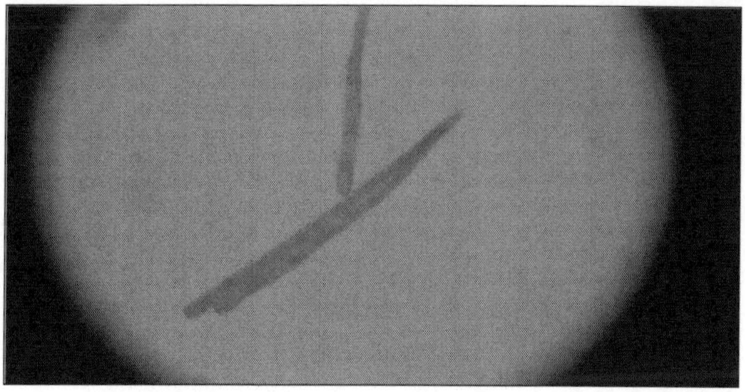

Survival

Internally infected seed

Spread

Seed and air

Environment

Cool and humid condition

Control

1. Seed treatment with systemic fungicides like bavistin 0.1 per cent
2. Seed treatment with thiram (0.2 per cent)
3. Two spray of Bordeaux mixture at one month intervals

Rust

Causal Organism

Uromyces anthyllidis

Symptoms

The yellow spots bearing aecia in clusters are earlier symptoms of the disease. The disease is observed in the form of minutes brown pustules on both surfaces of leaves. The disease is minor importance but under cool and wet climate in northern India entire plant may be damage with disease.

Control

1. Spray of fungicide mancozeb (0.2 per cent) and bayleton (0.5 per cent)
2. Adopt good cultural operation and field sanitation.

Other Disease

1. Collar rot–*Rhizoctonia solani* Kuhn.
2. Root rot–*Alternaria alternata* (Fr.) Kessiler
3. Yellovein viral symptoms

Downy Mildew

Uppal *et al.* (1935) reported the occurrence of downy mildew from Bombay for the first time and Sharma and Munjal (1977) recorded this disease from Himachal Pradesh. In the state, it was considered a minor disease as there were no reports of its occurrence in serious proportions. However, during last 3-4 years this disease is emerging as a serious problem in Solan district of Himachal.

Symptoms

The leaflets exhibit yellow patches on the upper surface and grayish growth on correspondingly lower surface. The grayish cottony growth was observed on lower surface of leaf. Typical symptoms of downy mildew were observed on fenugreek plants. The adaxial surfaces of the leaves exhibited small chlorotic spots often at

the leaf margins, while the abaxial surfaces exhibited a grayish violet, felty growth. The disease also caused stunted growth of the plant. Disease caused heavy yield loss when occurred in early stage.

Pathogen

Grayish cottony growth consist conidia and conidiophores of the pathogen. Conidiophores formed on the underside of the leaves, conidiophores are branched dichotomously 6 to 10 times, terminally forked and measured 274.30 to 538.65 µm (average 406.47 µm) with slightly swollen bases (7.5 to 10 µm broad which bears conidia at tip of branching. Conidia were slightly pigmented, oblong to ellipsoid and measured 26.64 to 34.97 x 19.98 to 26.64 µm (average 29.82 x 22.76 µm), on the basis of the morpho-taxonomic characters of conidia and conidiophores, the pathogen was identified as *Peronospora trigonellae* Gaum.

Survival

Oospores are formed towards maturity which is responsible for survival of pathogen.

Spread

Oospores and conidia

Environment

Oospores are favoured 24-27°C temperature and at least 40 per cent relative humidity and conidia formation is favoured 18°C temperature.

Control

Resistant Variet Hissar Mukta, Hissar suvarne

Damping Off

Symptoms

The infected seedlings appear water soaked discolouration and soft emitting a bad odour. The infection mostly occurs at or below the ground level and seedlings topples down and withers off

Causal Organism

Pythium aphanidermatum

Control

1. Spray carbendazim@ 1.0 g/litre water
2. Fallow cultural operation and crop rotation
3. Seed treatment @3.0g/kg seed with benomyl or carbendazim
4. Soil amendments like neem cake @10.0q/ha and FYM @ 10-25t/ha
5. Deep summer ploughing and crop rotation

Chapter 17

Cumin (*Cuminum cyminum* Linn.)

Powdery Mildew

Powdery mildew is a serious disease of Maharastra and Rajasthan. It is appear late in the season and reached maximum intensity at the time of flowering and seed formation.

Symptoms

Small white or grayish specks appear on the lower leaves, they gradually enlarge and coalesce covering entire leaf surface with mycelia growth and conidia of the fungus. In severe infection, a leaf show an ashy, white appearance and rapidly proceeds to stem flowers and fruits. The proper setting and development of fruit s are not taking place consequently there may be heavy reduction in yield. Infected field appear to be dusted with white flour. Under dry condition spread of disease is checked. Small white or grayish specks appear on the lower leaves, they gradually enlarge and coalesce, covering entire surface with mycelia growth and conidia of the fungus.

Causal Organism

Erysiphae polygoni DC.

Mycelium of fungus is entirely superficial. Only houstoria is sending into the epidermal cell. Hyphae are septate branched, white tangled mass and formed appresoria on the wall of epidermis. Conidiophores arise from upper surface of the mycelia growth; they are erect, stout bearing several conidia in chain. Conidia are unicellular, hyaline, and elliptical and barrel shaped, Germination takes place in light water by producing germ tube. Fungus also produce ascospores in cleistothecia which remain survive in unfavorable environmental condition.

Survival and Spread

Fungus survives as dormant mycelium on seed and disease is carried over from one season to another by infected seed. Secondary infection takes place by conidia.

Environment

Warm humid environment, Disease usually assumes serious form when maximum temperature reaches 26.5°C, below 25°C is not favour the disease. Rainfall is not necessary but light showers of rain increased the intensity, humidity is provided by irrigation water which favors the disease.

Control

1. Disease can be effectively checked by one application of sulphur dust @ 28 kg/ha at the time of flowering or disease initiation; if required, second application @ 13.5 kg/ha at the time of seed formation.
2. Application of perenox (Copper fungicide) completely control the disease, it should be applied before the outbreak of the epiphytotic
3. Karathan, Cosan and cumin @ 0.1 per cent are also effective against powdery mildew.
4. Use resistant variety.
5. Seed treatment with thiram @ 0.25 kg/100 kg seed give infection free planting materials.

Blight

Blight disease was first time reported from Bombay (Mumbai) by Uppal *et al.*, 1935. It is wide spread disease occurring all cumin growing area of Rajasthan and Gujarat.

Symptoms

The infected plants show whitish dark necrotic areas on the arial parts of plant mainly on tip and margin of young leaves. They enlarge and coalesces the spot and turn purple brown which finally shows black blighting appearance on the leaves, stem and blossoms resulting the killing of the infected plants. Seed of infected plant become shriveled, dark coloured and non-viable.

Causal Organism

Alternaria burnsii

Survival

Plant debris and collateral host

Dispersal

Seed and air

Control

1. Seed treatment with Agallol, Aureofungin, Ceresan or difoletan
2. Four Spray of Blitox or capton or Mancozeb @0.2 per cent at 10 to 20 days intervals
3. Prophylectic spray of 0.2 per cent dithane M-45 or Difolatan or Dithane Z-78 at 60-90 days 4-5 spray at 10 days intervals

Wilt

Symptoms

It is destructive disease of cumin crop and affected all growth stage. The severity was observed mare at early stage of growth when plant reached 2-5 cm height. The infected plant exhibited drooping of leaves and epinasty later lead to death of the plant. Such plant can pulled out easily from the soil. The wilt infection may occur in patches at any stage of crop growth. Roots of the diseases plants bear dark brown marking. After occurrence of the wilt in cumin it becomes very difficult to reduce the damage.

Causal Organism

Fusarium oxysporum f.sp. *cumini*

Survival

Soil and seed

Dispersal

Soil and seed

Control

1. Summer ploughing should be done

2. Crop rotation at least 4 years with mustard and pearl millet
3. Healthy and disease free seed should be used
4. 0.2 per cent capton shoud be used for soil drenching
5. Seed treatment can be done with Trichoderma viride or T. harzianum @ 4g/kg seed
6. Soil solarization of infected field during may-june with white transparent ploythine sheet.

Chapter 18

Cinnamon (*Cinnamonum zeylanicum*, Blume)

Bark Canker

The disease was originally described by Rands (1922) from Indonesia (Sumatra). It has caused a bark canker of Cinnamon. The pathogen has also reported in India (Rangaswami, 1961)

Symptoms

The affected plants shows irregular, vertical stripes of dead bark,1-5 cm wide, beginning at or below soil level and several meters long. The canker is sunken and at the advancing upper edge drops of amber or wine coloured exudates appear. A narrow, black line divides the healthy and diseased tissue.

Causal Organism

Phytophthora cinnamomi, Rands

It is worldwide in distribution coralloid becoming broad (8μm) and very tough hyphal swelling typically spherical and in cluster sporangiophores thin occasionally branched, more often proliferate through the empty sporangium. Sporangium which are formed in aqeous solution are broadly ellipsoidal or ovoid, no papilla, sex organ rarely produced in single culture, but abundant when grown *with phytophthora cryptogea* or an opposite strain of *P. cinnamomi*. Oogonia is 40μm, wall smooth, becoming yellow or golden with age. Antheridia are always amphigenous, long 21-23μm x 17μm. Oospores filling in the oogonium colourless thick. Optimum temperature for growth in culture media is 24-28°C and same for *in vivo* also.

Control

1. Soil application of sulphur is recommended for control.

Leaf Spot and Die Back

Leaf spot and die back diseases is caused by *Colletotrichum gloeosporioides.*

Symptoms

Small deep brown specks appear on the leaf lamina, which later coalesce to form irregular patches. In some time the affected portions are shed leaving shot holes on the leaves and late the entire lamina is affected and the infection spreads to the stem causing die back.

Control

1. Pruning the infected branches
2. Spray Bordeaux mixture @ 1.0 per cent

Seedling Blight

Symptoms

Seedling blight occurs in nursery. The fungus causes light brown patches which girdle the stem resulting in mortality of seedlings.

Causal Organism

Diplodia sp.

Control

1. The disease can be control by spraying of Bordeaux mixture @1. per cent

Chapter 19

Clove (*Syzygium aromaticum* L.)

Clove is the second most important spice of the world as judges from the world next to black pepper. In India it is confined to Kerala and adjoining area of Tamil Nadu.

Leaf Spot, Twig Blight and Flower Shedding

Colletotrichum gloeosporioides Penzig

*Colletotrichum l*eaf spot of clove was reported from Kerala by Jose and pailey (1966). Disease caused severe twig blight and flower shedding besides leaf spot. It was found to appear in three distinct phase namely leaf spot, twig blight and flower shedding.

On the leaves the disease manifested as small circular to oval- brown specks scattered all over the lamina. The speck enlarges and develops all over the lamina as spots with ashy-grey centre and darker margins. Spot coalescence and form necrotic patches. Then infection extend on young twigs initially appears in the form of brownish dots, which enlarge and show die-back symptoms with eventually death of twigs. The infection also notices on flower which develops in serious proportions during periods of heavy and continuous rainfall. The first visible symptoms appear as blackening of flower buds. Immature buds are more susceptible to infection. Whitish growth of pathogen can be noticed on fallen flower buds.

Causal Organism

Colletotrichum gloeosporioides Penzig

Perfect Stage: Glomerella cingulata

Acervuli on necrotic areas are clearly defined lesion on any part of the host. Setae are globous rounded, elongated or irregular in shape, setae variable in length, rarely 200µm x 4-8µm wide with 1-4 septa, brown slightly swollen at the base and tapered to the apex on which conidia are occasionally borne. Conidia cylindrical with otuse end, some time slightly ellipsoid with rounded apex and narrow truncated base. Hyphae aseptate, uninucleate formed on aseptate hyaline or faintly brown cylindrical phialidic conidiophores. Mycelium of the fungus in culture is grayish-white, thin and sparsely septate. Conidia are more variable in shape and size than on the host weed and fallen leaves, twigs and flower are the source of infection.

Control

No control measure is suggested. Prophylactic spraying of the fungicide are keep the pathogen under check.

Seedling Wilt

Seedling wilt is a serious problem in nurseries. The leaves of affected seedlings lose their natural luster, droop and ultimately die. The root system and collar region of the seedlings show varying degrees of discolorations and decay.

Causal Organism

Cylindrocladium sp., *Fusarium* sp. and *Rhizoctonia* sp.

Control

1. Remove the affected plants
2. Treat with carbendazim 0.1 per cent both as spray and soil drench or copper oxychloride @ 0.2 per cent

Leaf Rot

Leaf rot is characterized by dark brown spots with a yellow halo on leaves and is caused by Colletotrichum gloeosporioides. Spots appear on the buds resulting in their shedding.

Control

Prophylactic spray of Bordeaux mixture @1.0 per cent.

Little Leaf

This disease has caused by virus and was found in 1955-56 in Kanyakumari district of Tamil Nadu.

Symptoms

The affected tree exhibited complete repression of all tree and vegetative characters. The trunk was observed with numerous protuberances. In acute stage trunk appear cane shaped, leaves were reduced in size from the base to the apex of a shoot. Internodes shortened, primary and secondary branches were also arranged closer and become shorter towards the periphery. The growth of the trees arrested. This disease mostly found in nursery where all above symptoms appears. The disease is not transmitted by sap inoculation or by grafting.

Control

200 ppm of tetracycline spray at five times at biweekly intervals is found effective.

Chapter 20

Coriander (*Coriandrum sativum* L.)

Stem Gall

The stem gall is a common disease of coriander in northern part of India. It is causing heavy yield loss. The loss may be upto 23-60 per cent.Disease has been reported to be severe form near in Gwalior district (MP) in 1952-53 and in Uttar Pradesh (1984).

Symptoms

The disease appears as a tumor like growth on leaf veins, leaf stalks, peduncles, stem and fruits. The swelling on the veins give hanging appearance to the leaves. The enlarging swelling may attain a size of about 2.0 cm long and 0.5-1.0 cm broad. The disease infected plants remain stunted and when the plants inflorescence affected, seed production may be reduced considerably. The pathological anatomy of floral parts and fruits revealed that partial infected seeds are viable. Chlamydospores of the pathogen are presented in the affected parts. The gall develops as a chemically

induced swelling, arising from the surface of the leaf lamina, veins, mid-ribs and petiole. On the leaf lamina it forms yellowish budges standing out on the upper surface, but missing the black sori that are typical of an otherwise similar gall caused by the rust fungi *Puccinia aegopodii*. On veins, mid ribs and petioles the gall appears as translucent yellow-white swellings that are often elongated as blister-like.

Causal Organism

The stem gall disease is caused by *Protomyces macrospores* unger. The fungal mycelium is found only in tumors. The hyphae are intercellular and septate with irregular branching. The scattered cells in the hyphae swell up to produce ellipsoid or globose bodies converted in to Chlamydospores. The Chlamydospores are smooth, three layered walls, ellipsoid to globose measuring 50-60μm in diameter. They function as the resting spore of the pathogen and germinate by rupturing outer wall in the presence of water. A vesicle is produced through rupturing wall of chlamydospore. The cell content of chlamydospore move in to vesicle and the nuclei present in it divide several time and produce 100-200 daughter nuclei which further divide meiotically and each produce four haploid nuclei. The each haploid nuclei produce a spores inside the vesicle which bursts and releasing the spores to germinate and infect the host plants

Survival and Spread

The fungal pathogen can survive in the soil in the form of Chlamydospores which work as primary source of disease inoculums. The pH range between 4.6-7.4, high soil moisture and shade predispose the plant to fungal infection. Microclimate

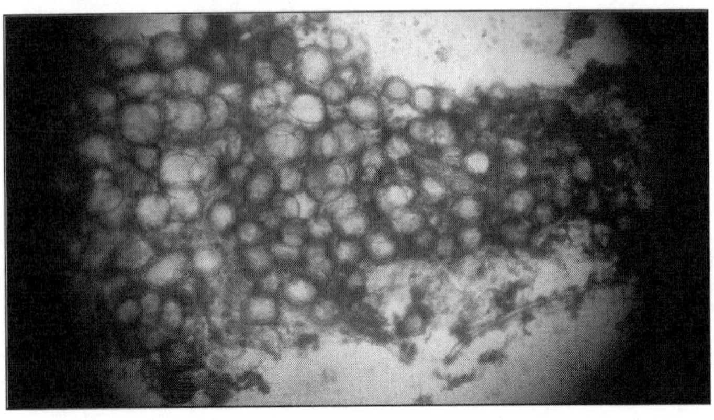

and stage of development of the crop also affect to disease development. Low temperature and abundant dew with slightly higher day temperature seemed to predispose plants to attack. Disease also increased in field by heavy irrigation, poorly drained soil in low lying and shady areas

Control

1. Use of clean and healthy seeds
2. Proper field sanitation
3. Removal and destruction of the diseased plants
4. Crop rotation.
5. Treatment of soil and seed with Thiram, Hexaconazole and propiconazole @ 0.2 per cent.
6. Sulphathiazole and sulphadiazine (250-1000ppm) inhibited the germination of chlamydospore. Complite inhibition of chlamydospore germination by Streptomycin, Agrosan and pennox.
7. Foliar spray with Captafol (0.2 per cent) or Corbaxil (0.1 per cent) at 15 days interval
8. RCr-41 and K-selection are resistant to stem gall.

Other Minor Disease

1. Ramularia blight: *Ramularia corianderii*
2. Root rot: *Macrophomina phaseoli*
3. Stem rot: *Phoma applicata*
4. Leaf spot: *Cercospora corianderii*
5. Leaf blight: *Alternaria alternata*
6. Damping off: *Pythium irregular*

Powdery Mildew

Symptoms

White circular patches on young parts of stem and leaves are appeared during February and March. These patches increase in size and coalesces which cover large area on leaf surface. Infected leaves reduced and flower drop, seed formation very less, which formed small and shrivelled seeds. 20-30 per cent less seeds are estimated.

Causal Organism

Erysiphe polygoni

Dispersal

Air borne

The disease generally appears in early February to March in Uttar Pradesh, Madhya Pradesh, Bihar and Rajasthan and other parts of the country according to the prevailing weather conditions.

Enviroment

Warm moist conditions are favourable for spread of the disease.

Control

1. Early planting
2. Seed treatment with thiram
3. Dusting of sulphur is found effective for disease control.
4. Spray of wettable sulphur or sulfex @0.6 per cent is also effective management practices
5. Resistant cultivars are CS-362, CH-15 NDCor-68 and CS-52

Wilt

Symptoms

Wilt disease is observed at all stages of plant growth. The severity of infection are more at the early stage of plant. The disease is easily recognized in the field by dropping of the terminal portion followed by weathering and drying of plant. Partial wilting is also observed where plant growth is arrested. The plant leaves become pinkish yellow in colour tissues of vascular bundle discoloured due to growth of pathogen. Yield losses are observed high when disease occurred in early stage and low in later stage.

Causal Organism

Fusarium oxysporum f.sp. corianderii

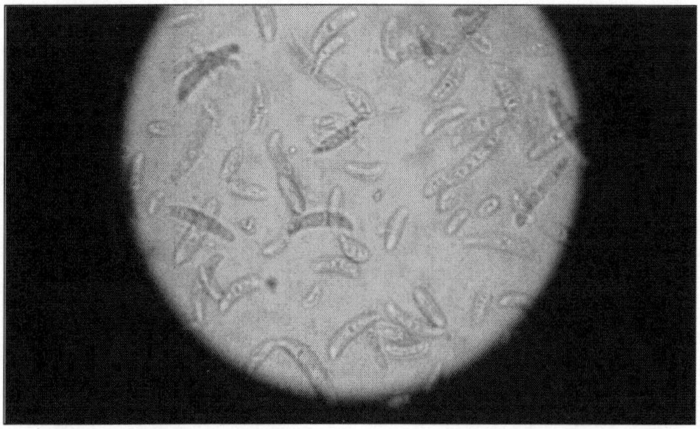

Survival

Soil and plant diberis

Dispersal

Soil borne

Environment

High soil moisture (more than 70 per cent)

Control

1. Crop rotation
2. Crop sown first and second week of November
3. Adjust soil pH at 8.2
4. Neem cake amendment in soil is also found effective

Soft Rot

Symptoms

Soft rot symptoms appear on leaves. Leaves become sudden yellow and yellowing spread all parts of the plant. In severe case all infected plan die.

Causal Organism

Erwinia aroiatae or E. carotovora

Survival

Seed

Dispersal

Seed and air

Control

1. Seed treatment with @ 0.4 per cent formalin, or 0.1 per cent panogen
2. Seed treatment with dry hat air treatment at 53°C for 30 minutes

Chapter 21

Black Pepper (*Piper nigrum*)

Foot Rot

Symptoms

All part of vine are vulnerable to the disease and the expression of symptoms depend upon the plant part infected and the extent of damage. The symptoms of diseases are noticed as one or more black spots appearing on the leaves which have characteristics fine fibre like projection on the leaves, it spread toward the margin and rapidly defoliate the plants. The tender leaves and succulent shoot tips of freshly emerging runner shoots trailing on the soil, turn black when infection advances. The disease spread to the entire vines from these infected runner shoots and leaves during intermittent showers due to rain splash.

If the main stem at the collar region get infected and damaged then entire vine wilted followed by shedding of leaves and spikes with or without black spots. The

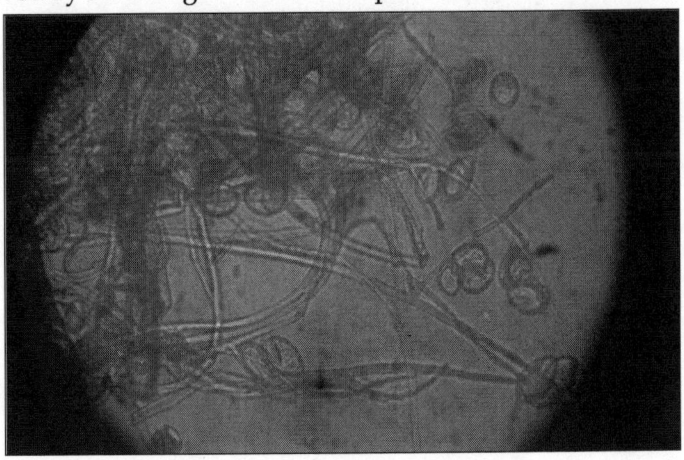

branches break up at nodes and entire vine collapses with in month. This symptoms appeare during October-November and onwards. These vine may recover after rains and survive for more than two season

Causal Organism

Phytophthora capsici

Control

1. Removal and destruction of dead vines along with root system
2. Planting materials should be disease free plants
3. Nursery bed should be fumigate or soil solarized
4. Adequate drainage should be done
5. Soil drench at the radius of 45-50 cm with copper Oxychloride @ 0.2 per cent or Ridomil or mancozeb during May-June and August –September and October
6. Foliar spray with Boreaux mixture @ 1.0 per cent.

Anthracnose

Symptoms

The disease appears towards the end of the monsoon. The affected berries show brown sunken patches during early stages. Later stages discolouration increases and berries shows cross splitting and gradually turn black and dry. The irregular brownish halo lesion also seen on the leaves.

Causal Organism

Colletotrichum gloeosporioides

Control

1. Spray Bordeaux mixture @1.0 per cent.

Spike Shedding

Spike shedding is a emerging and serious disease of all pepper growing area. This occurs during June and July. In serious condition delayed the flowering or drooping the flower.

Control

1. Irrigation done properly during March to June.
2. Spray prophylactic spray of Bordeaux mixture.

Chapter 12

Cardamom (*Elettoria cardamommum*)

Kattee or Marble Mosaic Disease

Causal Organism

Virus

Survival and Spread

Insect transmitted, survive in rhizomes

Symptoms

Chlorotic symptoms running to paralal to veins on young leaves, affected plant deteriorate rapidly, leaves reduced in size, marked drawfing and stunting of stools, fruiting suppressed.

Control

Use diseases free planting materials for plantation, Destroy the infected clumps from field, Prophylactic spray with insecticides.

Rhizome Rot

Symptoms

Pale and yellowish leaf seen on infected plants and leaf die premature ty. Letter decay and rotting occurred at the base of stem. Rhizome growth of plant stunted, collapse and pulled out easily.

Cousal Organisms

Pythium aphanidermatum, P. vaxan and *Cephalosporium* spp.

Survival

Soil and rhizome

Spread

Soil borne and rhizome

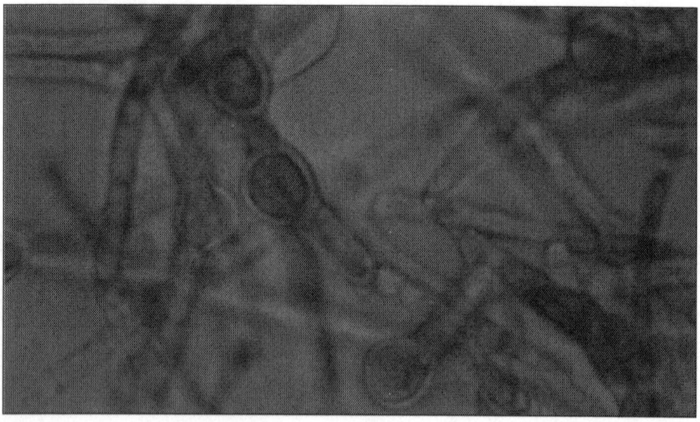

Environment

Poor drainage soil and high soil moisture

Control

1. Infected clum should be destroyed
2. Soil application of ammonium phosphate or calcium superphasphate@55-60g/clump

3. Good drainage systems
4. Drenching of Bordeaux mixture@1 per cent

Chirke (Mosaic Streak)

Survival and Spread

Insect transmitted, Survive in rhizomes

Symptoms

Mosaic streaks on leaves, streaks coalesce and leaves dry up

Leaf Rot

Coniothyrium sp.

Survival and Spread

Spread by wind borne spores

Symptoms

This disease is found in rursery, water soaked lesions develped on leaves which spread rapidly to cover entire blade. Rots and rotting spread in entire shoot. In case of severe infection whole nursery may be destroyed.

Control

Prophylactic sprays give with copper fungicides 3-4 times at 15 days intervals during monsoon season.

Chapter 23

Celery (*Apium graveolens* L.)

Celery is growing in India since 1930. It is mainly growing in Punjab, Haryana and Uttar Pradesh. The crop is affected by several disease that is-

Stem Rot

Symptoms

Black spots are formed on the collar region of the plant which coalesce and infect to the part of stem and turn water soaked lesion that cause to rotting symptoms. In later stage whole plant become dry.

Causal Organism

Rhizoctonia solani, Perfect stage *Pellicularia filamentosa*

Cercospora Blight

Symptoms

Small circular brownish yellow spots are formed on the foliage. Numbers of conidia are formed on lower surface of leaves. High moisture condition with droplets of the water increase the germination percentage of spore on foliage.

Causal Organism

Cercospora apii

Survival

Seed borne

Dispersal

Primary infection by seed and secondary infection through conidia by air.

Environment

High moisture and heavy persistant of dew is favour disease development.

Septoria Blight

Symptoms

The small, circular water soaked spots are formed on the leaves which later coalesce and give blight appearance. The disease starts from outer lower leaves and spread upward direction. In severe case affected plant become defoliated.

Causal Organism

Septoria apii

The blackish pycnidia are formed in the spots, they are enlarge number. The spores are straight, 0-4 septate and are ejected out in a cirrhus. Pycnidia are embedded in seed coat.

Survival

Seed and plant refuge in soil

Dispersal

Seed and soil

Control

1. Disease free Seed
2. Seed treatment with fungicide

Vascular Wilt and Fusarium Yellowing

Symptoms

The disease appears at seedling stage, if the conditions are warm, the seedling become dry. The remaining plant stunted and leaves exhibit yellowishness and splitting. The roots are injured and exhibit brown discolouration in the vascular stand.

Causal Organism

Fusarium oxysporium f.sp. apii

Survival

Seed as well as soil

Dispersal

Seed and soil

Bacteria Leaf Spot

Circular or angular spots are formed which are yellowish first and later turn brown with yellow halo. In Severe case plant give rise blight appearance and in later case it dies.

Causal Organism

Pseudomonas syringae pv. Apii

Survival

Seed and soil

Dispersal

Rain splash

Control

Spray copper fungicides

Chapter 24

Ajawain (*Trachyspermum ammi* L.)

Ajwain is an important seed spices with excellent medicinal properties used in Kichen as well as making pickles. It is grown in Rajasthan, Gujarat, MP, UP, AP, Tamil Nadu and West Bengal. The crop is suffering many diseases which reduces productivity of crop.

Leaf Blight

Initially symptom appears on the first leaves of plant as small ring like spot which coalesces and develop blight symptoms. The tender stems petioles and leaves are badly affected. It is small irregular radish brown with gray centre.

Causal Organism

Alternaria alternate

Survival

Weed and other host plant

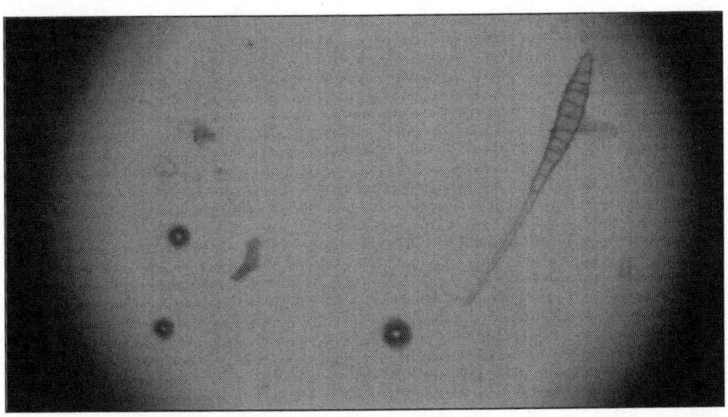

Dispersal

Air and seed

Control

Spray mancozeb @0.25 per cent two to three spray at 10.15 days intervals

Collar Rot

Symptoms

Disease lesion formed on the stem near the soil surface and white hyphae grow around the lesion which letter covered the lower portion of the stem. The leaves exhibited yellowing from the base of stem and develop in patches. In severe case plant dies.

Causal Organism

Sclerotium rolfsii

Survival

Soil and plant debris

Dispersal

Soil

Control

Summer plaughing and crop rotation

Powdery Mildew

All the above ground parts are infected and covered the white powdery mass. When the inflorescence gets infected all the flowery aborted and no seed formation are found.

Causal Organism

Erysiphae polygoni

Survival

Seed and plant debris

Dispersal

Seed and air

Control

Spray of wet table sulphar @ 0.25 per cent or Karathene @ 0.1 per cent.

Chapter 25

Dill or Sowa (*Anethum suwa* or *A. graveolens* L.)

Tender leaves of sowa are used as vegetable and seed spices Two spices are grown in India mostly in Rajasthan, Gujarat, MP, UP, AP, Tamil Nadu and West Bengal. It is affected by several diseases causing fungi bacteria and virus.

Wilt or Root rot

Plant is affected at seedling stage and shows yellowing symptoms at letter stage of crop growth, leaves of plant become yellowish, droop then wilted and finally die because vascular bundle of root split and become brown and discoloured.

Causal Organism

Fusarium oxysporum

Survival

Soil, seed and Plant debris

Dispersal

Soil and seed

Control

1. Seed treatment with mercurial fungicides
2. Summer ploughing and long crop rotation

Powdery Mildew

Symptoms

White powdery floury mass appear on foliage, stem branches, flower and fruit which have consist conidia of fungus. In severe infection flower droop and low fruit setting have been observed. The yield and quality of seeds are badly affected.

Causal Organism

Erysiphe polygoni

Survival

Seed and plant debris

Dispersal

Seed and air

Environment

Warm and humid weather

Control

Wettable sulphar @0.25 per cent or Sulfex 0.2 per cent spray at 15-20 days intervals

Nigella or Kalonji (*Nigella sativa* L.)

Nigella is native of Mediterranean region. In India it is found in wild and commercially grown in Punjab, MP, UP, Bihar and Assam. The seed of Nigella contain 0.7 per cent essential oil which is known as nigellone used in cough and asthma. The most important disease of Nigella is root rot.

Root Rot

Symptoms

The roots are attacked by fungus which intrupt the vascular boundle that causes to rotting symptoms in tender roots and get killed. In later stage main root shriveled and whole plant become dry.

Causal Organism

Rhizoctonia solani, Perfect stage *Pellicularia filamentosa*

Survival

Soil and plant debris

Dispersal

Soil and equipment

Control

Crop rotation and summer ploughing

Anise (*Pimpenella anisum* L.)

Anise is commercially grown in India. It is a crop of temperate and subtropical region. Anise have carminative, digestive, antispasmodic soporific and antiseptive properties. The crop is suffered by several disease, the most important disease is Alternaria blight.

Alternaria Blight

Symptoms

Small brownish spots develop on the leaves, stem branch and fruits which enlarge and coalesce to give blight appearance. In letar case spot becme blackish, conidia and conidiophores formed inside the spot.

Causal Organism

Alternaria alternate

Survival

Weeds and collateral plant

Dispersal

Air

Control

Spray Dithane M-45 or Copper oxychloride @ 0.25 per cent

Diseases of Plantation Crops

Chapter 26

Coffee (*Coffea arabica* or *C. robusta*)

Rust of Coffee

Causal Organism

Hemielia vestatrix

The mycelium of fungus is intercellular with globose houstoria. The sori are hypophyllus densely scattered and give a podery appearance on yellowish orange rounded blotches on the leaves. These sori consist numerous hyphae and bearing clavate filaments emerging through the stomata. The tip of filaments bear numerous uredospores The uredospores are reniform and micro in size. Teliospore is formed in same sori. They are one celled and pedicellate and turnip shape. They have no dormancy period. Pycnial and aecial stages of this pathogen have not been reported.

Survival and Mode Infection

Pathogen survive in uredial stage.

Environment

High moisture and high temperature with intermittent rain favour the disease development.

Symptoms

Pathogen mostly confined to the undersurface of leaf blade in the formed of small yellowish spots. Spots enlarge, colour turns reddish brown and covered with powdery masses of spores, upper surface of leaf corresponding to spots shows chlorotic lesions, host tissue dries, leaf drop, and defoliation causes reduction in yield.

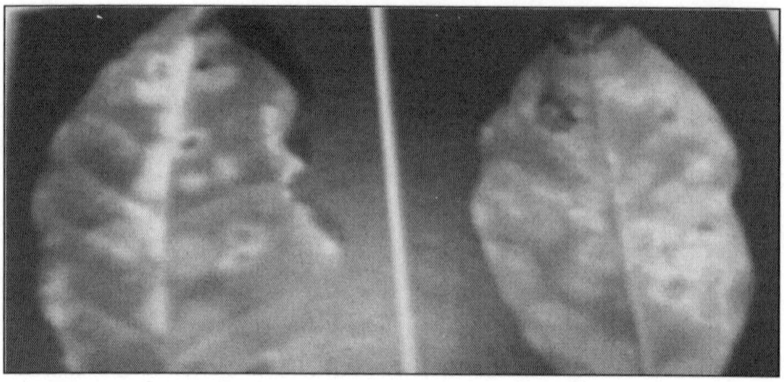

Control

1. Spray copper oxychloride or dithiocarbamate at 10-15 days intervals.
2. Systemic fungicide like bavistin can be used.
3. Grow resistant varieties.

Kolerega Disease

Causal Organism

Corticium koleroga

Survival and Spread

Survive on infected twigs, leaves, saprophytically on dead plant debris and collateral hosts.

Symptoms

Leaves and berries commonly affected. Leaves and berries membranous become grayish white coating which easily peeled off, affected plant parts turn brown, during wet weather infected parts rotted and caused heavy yield loss.

Control

1. Shading of plant should not be allowed
2. Destruction of infected plant parts
3. Periodic spraying with 1 per cent Boredeax mixture

Die Back or Anthracnose

Causal Organism

Colletotrichum coffeanum

The mycelium are consist narrow septate hyphae. Aservuli are formed abundantly on the affected host surface. Conidiophores are developed in aservuli. At the apex of conidiophores conidia are formed. Setae are common in the ascervuli

Survival and Spread

Spread by air born conidia, survive on infected plant parts and debries.

Symptoms

Round, irregular, grayish spots on leaves and stem are seen; on the berries dark coloured depressed spots are formed. When the pathogen attack on the twigs shows die back symptoms.

Control

1. Provide good drainage.
2. Spray bavistin or propiconazole regularly

Chapter 27

Tea (*Camelia sinensis*)

Blister Blight

Causal Organism

Exobasidium vexans

They are typically clubbed shaped, non septate basidium. Basidia are formed directly from dikaryotic hyphae between the epidermal cells of host, push through the cuticle and form a layer on the surface of the host, which usually bear four basidiospore on sterigmata. Basidiocarp lacking, hymenium covering the surface of parasitizing plant tissue. The basidiospores are uninucleate. They germinate either by budding producing blastospores or by germinate forming germ tube which infect the host and develops into a haploid primary mycelium.

Survival and Spread

Endemic in plantation crop and found round the year, basidiospores disseminated by wind.

Symptoms

Small yellowish spots formed on the leaves, upper surface develops a shallow depression, Crosponding to the depression the lower surface bulges into blister like swelling, much curling and destoration of leaves, tender shot also affected and showing swelling and twisting.

Control

1. Prophylactic sprays with copper fungicides
2. Picking and destroying older infected leaves

Red Rust

Causal Organism

Cephaleurous parasitica and C. *mycoidea*

Survival and Spread

Endemic in plantation crop

Symptoms

Small redish brown disc like spots and yellow hair appear on the leaves and green twigs, twigs harden prematurely forming scaby patches which crack.

Control

Improve general cultivation and manuring, Spray 1per cent Borredeax mixture.

Brown Blight

Causal Organism

Colletotrichum camelliae

Survival and Spread

Survive in infected plant debris, spread by air borne conidia

Symptoms

Yellow brown spots on upper surface of leaf, later appearing on lower surface, colour deepens to chocolate brown, affected tissue die, severe infection cause defoliation

Control

Prophylactic spraying with copper fungicides

Leaf spot

Causal Organism

Cercospora theae

Survival and Spread

Survive in infected plant debris, spread by air borne conidia

Symptoms

Characteristic spots on older leaves which latter formed irregular patches of dead tissue, in severe cases leaves fall and defoliate.

Chapter 28

Rubber (*Hevea brasiliensis*)

Abnormal Leaf Fall and Stem Rot

Causal Organism

Phytophthora palmivora

Survival and Spread

Dormant inside bark or in soil, Spread by wind borne sporangia.

Symptoms

Several type of symptoms produce on leaves dull grey spots later becoming irregular and form large patches, infected leaves fall off prematurely, tender shoots rot show die-back symptoms, when tapping panels and collar regions are infected the bark swells and bursts and amber coloured liquid ozez out, when bark removed there is dark discoloration of wood, fruit rot also occurs.

Control

Prophylactic sprays with 1 per cent Bordeaux mixture or other copper fungicides, spray of oil based copper oxychloride preparations, in bark infections, infected part scrapped washed with ceresin *etc.* and pasted with fungicidal paste.

Powdery Mildew

Causal Organism

Oidium heveae

Survival and Spread

Survival as dormants mycelium inside the bark, spread by wind borne conidia

Symptoms

Ashy powdery coating on tender leaves and shoot, tender leaves curl, crinkle and soon fall off and die- back symptoms appear.

Control

Dusting 325 mesh sulphur 10-12 kg/hectare 4-6 times, wettable sulphur 1.0 kg/400 liter water/ha can also be sprayed.

Chpater 29

Arecanut (*Areca catechu*)

Mahali or Koleroga or Fruit Rot/Nut Fall Bud Rot

Causal Organism

Phytophthora palmivora

Survival and Spread

Survive in plant debris and soil, spread by wind- borne spores

Symptoms

Tender nuts rot and fall off, buds also rot, fallen nuts are discoloured and covered with a whitish felt of fungal mycelium as in coconut palm

Control

Spray on bunch and crown with 1 per cent Bordeaux mixture or any other suitable fungicide

Stem Bleding

Causal Organism

Ceratostomela paradoxa

Survival and Spread

Pathogen enter through wounds. It is either cracks or incidental wounds.

Symptoms

Reddish brown liquid oozes develop on infected stem. The ooze dry and turns black, growth of plants is arrested and yield reduced, finally the tree wilt.

Control

Cut and scoop out discoloured tissues, paint the cut surface with hot coaltar or Bordeaux paint.

Foot Rot

Causal Organism

Ganoderma lucidum

Survival and Spread

Soil borne

Symptoms

Roots are first infected and in weak plants fungus spread upper part upwards killing entire root system, the disease plant die slowly.

Control

Infected plant digout and burn, Spread 1-13 kg sulphur and incorporate it in to the soil.

Chapter 30

Coconut (*Coccus nucifera*)

Phytoplasmal Diseases of Coconut Palm

Three important diseases of phytoplasmal origin have been reported. In India, root wilt and tatipaka are commonly found.

(i) Root Wilt or Kerala Wilt

Symptoms

The symptoms of the disease normally appear when the plants are about 30 months old. The most characteristic symptom is bending of leaflets called as 'flexidity'. In older palms, yellowing and marginal necrosis of the older leaves also develops. Parts of affected palms show degeneration of phloem, disorganized tracheal elements and tylosis in the meta-xylem. They eventually rot. The disease is not lethal but productivity is significantly reduced.

Pathogen

Mycoplasma-like Organism (MLO).

The disease is now known to be caused by a phytoplasma, a mollecute, formerly known as Mycoplasma-like Organism (MLO). The pathogen has pleomorphic bodies which can only be viewed in electron microscope. These organisms are limited to the sieve tubes of the phloem tissues. The root wilt phytoplasma is transmitted by insect vectors, *Stephanitis typica*-a lace-bug but *Proutista moesta* is the efficient vector.

Disease Cycle

The inoculums is available in the infected plants from where the pathogen spreads to new plants.

Control

Tetracycline therapy gives temporary remission of disease symptoms but no control of the pathogen.Movement of planting material from infested areas to new areas should be avoided.

(ii) Tatipaka Disease

Reduction in number and size of leaves, fasciations', light green leaves and chlorotic spots on leaves are the main symptoms of the disease. Abnormal bending of fronds, tapering of stem and production of small sized in fluorescence bearing atrophied empty nuts can also occur.

Chapter 31

Deficiency and Physiological Disorders

Tomato

Blossom End Rot

It is most common and destructive type of disorder of tomato. Brown coloured lesions appear at the blossom end while the fruits are green. At the point of attachment of senescent petals water soaked spots develop which enlarge in size. The disorder results in sunken, leathery and dark coloured development on the affected part of the fruit.

Causes

1. High soil moisture and high temperature.
2. Moisture stress conditions.
3. Application of nitrogen in abundant quantity.
4. Deficiency of calcium.

Control

1. Give light and frequent irrigations to maintain optimum soil moisture.
2. Apply recommended quantity of nitrogen.
3. Spray calcium chloride (0.5 per cent) at fruit development
4. Grow resistant varieties.

Fruit Cracking

The surface of mature green and ripe fruits cracks at the stem end. It is a common disorder observed during rainy season. The cracking is of two types

Radial Cracking

The cracking's start from the stem end and develop in ripe fruits. It is more serious.

Concentric Cracking

The cracking appears around the shoulder of the green fruits.

Causes

Long dry spell during rainy season, Moisture stress condition due to irregular irrigations, sudden fluctuation day and night temperature.

Causes

Exposure of fruits to sun light, Boron deficiency in the soil.

Control

1. Maintain optimum soil temperature through light irrigations
2. Avoid pruning and staking of plants during summer.
3. Grow resistant varieties like Sioux, Punjab chhuhara, Pusa Rubby, Roma. PantT-1, Arka Sourabh.
4. Spray Borax at 0.3-0.4 per cent two to three times.
5. Harvest the fruits before they ripe fully

Puffiness

When the fruit attain about two third of normal size, the growth of internal tissues retards whereas the outer wall develop normally. The affected fruits light in weight, lack of firmness and partially filled. The disorder is more common in winter season.

Causes

1. Lack of fertilization
2. Abortion of embryo after fertilization, necrosis of vascular and placental tissues of fruits, High temperature and high soil moisture
3. Application of high dose of nitrogen.

Control

1. Avoid over watering to crop.
2. Grow resistant varieties
3. Apply recommended dose of nitrogen

Cat Face

The affected fruits are characterized by the distortion of the blossom end. Rot and development of ridges, furrow indentations and blotches are common features of disease.

Cause

Unfabourable climatic condition during flowering cause distortions of growth of the pistal cells.

Control

1. Grow resistant and free from disorder varieties
2. Cultivate tomato in normal growing periods

Sun Scald

The fruits exposed to intense sun light, develop yellow or brown burnt patches. The affected tissues have blistered water soaked areas. The rapid desiccation of water produces sunken area, which has white or grey and yellowish coloured in green and red coloured fruits.

Control

1. Grow abundant foliage varieties
2. Raise the crop with high density.
3. Control the foliage diseases and insect pests timely which cause defoliation
4. While picking of fruits avoid the turning of plant
5. Avoid training and pruning of plant during summer months

Frost Injury

Tomato is highly sensitive to low temperature injury. The affected plants become darken, whither and desiccate. Flowers drop and results in poor fruit setting. The affected fruits become soft, water soaked and dull coloured.

Control

1. Grow low temperature tolerant varieties like Push sheeted
2. Cover the nursery bed with patches
3. Raise soil temperature through light and frequent irrigation during winter season.

Brinjal

Poor Fruit Set

On the basis of style length brinjal flower can be grouped into four categories

1. Long style with big ovary
2. Medium style with medium ovary
3. Pseudo-short styled with rudimentary ovary
4. True short styled with very rudiment ovary

Among these flowers only long and medium styled flowers produces fruits and pseudo- short and true short styled flower fail to set any fruits. Natural dropping of flowers and fruits is also a problem in brinjal. It varies from 30 per cent in long styled flower and 50 per cent in medium styled flowers

Control

Foliar application of 2,4-D (@2PPM) at flowering stage and NAA (60ppm) or PCPA 50PPM) at full blossom stage is recommended for high percentage of fruit set and early yields

Chilli

Blossom End Rot

It is most common and destructive type of disorder of chilli. Brown coloured lesions appear at the blossom end while the fruits are green. At the point of attachment of senescent petals water soaked spots develop which enlarge in size. The disorder results in sunken, leathery and dark coloured development on the affected part of the fruit.

Causes

1. High soil moisture and high temperature.
2. Moisture stress conditions.
3. Application of nitrogen in abundant quatity.
4. Deficiency of calcium.

Control

1. Give light and frequent irrigations to maintain optimum soil moisture.
2. Apply recommended quantity of nitrogen.
3. Spray calcium chloride (0.5 per cent) at fruit development
4. Grow resistant varieties.

Sun Scald

The fruits exposed to intense sun light develop yellow or brown burnt patches. The affected tissues have developed blister water soaked areas. The rapid desiccation of water produces sunken area, which has white or grey and yellowish coloured in green and red coloured fruits.

Control

1. Grow abundant foliage varieties
2. Raise the crop with high density.
3. Control the foliage diseases and insect pests timely which cause defoliation
4. While picking of fruits avoid the turning of plant
5. Avoid training and pruning of plant during summer months

Flower and Fruit Drop

It is a major limiting factor in chilli cultivation. Flower and fruit drop may be following reasons:

1. High temperature and low humidity at flowering cause excessive transpiration, which result abscission of buds, flowers and fruits
2. Low light intensity
3. Short day and high temperature

Control

1. Give light and frequent irrigations at flowering and fruit set stages
2. Spray the crop either NAA 50ppm or tricontanol 2 ppm at full bloom stage.

Potato

Black or Hollow Heart

In these disorder central tissues of the affected tubers shows dark grey or black discolorations due to sub-oxidation (Black heart). In advanced stage of disorder, the affected tissues dry and separate to from cavities brought about by very rapid growth of tubers (hollow heart). The symptoms are internal; therefore, it can be seen only after cutting the tuber. It is more common in large sized tubers.

Causes

1. High soil temperature (above 33° C) during growth and maturity of tubers.
2. Rapid growth and development of tubers followed by sudden irrigation.
3. Unfavorable supply of oxygen during storage and transport.
4. Storage of tubers at high temperature. Even 3-4°C temperature is not safe to store the potato.

Control

1. Follow close spacing, remove foliage and avoid over- fertilizers.
2. Irrigate the crop frequently to maintain soil temperature below 33°C.
3. Provide proper ventilation in storage and during transportation.
4. Storage of tubers in cold storage at 2.2-3.3°C temperature.

Greening

The tubers turn green in colour when exposed to direct sunlight. The green pigment produced is Solanin which is slightly poisonous and make the tubers unfit for consumption.

Control

Earthing up should be done to avoid the exposure of tubers to direct sunlight.

Sprouting in Storage

Sprouting of potato tubers in storage is the major problem which deteriorates the quality and make the product unfit for consumption. The intensity of sprouting depends on the variety, maturity, storage temperature and relative humidity.

Control

1. Grow resistant varieties which sprout late
2. Spray the crop with maleic hydrazide @ 3000 ppm at 15 days before harvesting
3. Store the potato tubers in cold storage 2-4°C and 90-95 per cent relative humidity

Cole Crops

Bolting

Premature formation of seed stalk without forming the head is known as bolting. It is commonly found in early cabbage and cause considerable losses to the farmers, bolting may be due to

1. Early sowing of seeds in the hot weather.
2. Presence of warm winter
3. Sudden and extreme change in temperature during crop growth.
4. Poor growth of seedlings due to inadequate nutrient supply, poor viability of seeds and attack of insect pests and diseases

Control

1. Grow bolting resistant varieties
2. Avoid the sowing of warm climate
3. Supply adequate amount of nutrients

Buttoning

Very small curd or buttons are formed. Leaves also remain small, which are enabling to cover the curd

Cause

1. Poor supply of N. Transplanting of over aged seedlings Sowing of early varieties
2. Transplanting of poor seedlings

Control

1. Right selection of variety
2. Apply recommended dose of N Transplant healthy and vigorous seedlings

Browning

At initial stage,water soaked areas develop in the centre of curds and in advance stage the stem becomes hollow with the cavity surrounded by water soaked tissues. Later curd changes to rusty brown and becomes bitter in taste

Causes

Deficiency of boron

Control

Apply borax @10-15 Kg/ha at field preparation or spray 0.1 per cent boric acid. Harvest the crop at proper maturity stage.

Whip Tail

It is characterized by the lack of leaf lamina and leaf consists of only bare mid rib which looks like whip tail.

Causes

Deficiency of molybdenum particularly in acidic soils having pH below 5.0.

Control

Apply 0.5-1.0Kg sodium or ammonium molybdate/ha at field preparation or spray the crop with 0.1-0.3 per cent ammonium molybdate with 0.1 per cent teepol as sticker. Add lime in the soil to raise the pH to 6.5.

Root Vegetables (Carrot, Radish *etc.*)

Heart Rot or Crown Rot

This disorder is caused by boron deficiency. The leaves dry in the crown which is covered with small deformed leaves. The older leaves wilt and become necrotic. The entire crown becomes necrotic and starts to decay. The inner portion of affected roots turns black and become unfit for consumption.

Control

1. Soil application of borax (10-15 Kg/ha) or foliar spray of boric acid (0.2 per cent) in 2-3 times at vegetative stage.
2. Avoid the sowing of beet in acidic soils
3. Avoid drought conditions by supplying regular irrigation

Speckled Yellows

It is due to deficiency of manganese. The leaves of affected plants show yellowish-green chlorotic mottled areas. The chlorotic areas become necrotic resulting to breaking of lamina. The leaf margins roll upward and turn into an arrow shaped outline which remains upright. Generally deficiency of Mn is observed in very sandy and very alkaline soils.

Control

1. Deficiency can be corrected by the soil application of manganese sulphate @ 5-10 Kg/ha or foliar spray of manganese sulphate (0.25 per cent) at 2 to 3 time.
2. Avoid planting on very sandy and alkaline soils.

Onion and Garlic

Sprouting of Bulbs

It is found both onion and garlic. Sprouting in white cultivars is more than in pink or purple cultivars. It is become due to excessive soil moisture at maturity and supply of nitrogen. Sprouting mostly occur in storage condition

Control

Sowing time adjust that harvest can be done in dry period. Withhold irrigation as soon as bulbs reaches to maturity, Apply less quantity of nitrogenous fertilizers, grow purple or pink coloured cultivars

Spray growth inhibitors like MH (2500-3000 ppm) 15-20 days before harvesting

Bolting

It is an emergence of seed stalk prior to time of their formation and adversely affects the formation and development of bulbs. Bolting is undesirable characters because it directly affects the bulb yield of onion

Causes

1. Rabi onion starts to bolt at 20-25°C temperature. Early and late transplanting induce bolting
2. Low temperature (10-12°C)
3. White cultivars are more sensitive to bolting
4. Poor supply of nitrogen in nursery and field

Control

1. Adjust time of transplanting
2. Grow non- bolting cultivars like Early Grano, Texas Early grano
3. Supply recommended dose of nitrogen

Pea

Marsh Spot

The deficiency of manganese in soil causes marsh spot of pea. Leaves of affected plants shows slight interveinal chlorosis. The flate surfaces of seeds have a brown spot or cavity in the centre and become hollow at later stages

Control

Application of Manganese sulphate @ 5-10 Kg/ha or foliar application @0.25 per cent twice or thrice.

Okra

Poor Seed Germination

Normally in north Indian plains, okra is sown in two seasons. In rainy season there is no problem of seed germination. However, prevailing of low temperature during February, which is an optimum time of sowing of spring summer crop, leads to poor seed germination.

Control

1. Seed soaking in fresh water for 24 hours and in hot water (45°C) for 30 minutes improve seed germination
2. Tie the water soaked seeds in a piece of cloth and keep into a heap of fresh cow dung overnight to stimulate germination.

Cucurbits

Excess of Staminate Flower and Low Fruit Set

Fruit set and yield depends on the number of pistillate and perfect flowers. Male flowers appear on the first 4-6 flowering nodes and later pistillate/hermaphrodite flowers. In the both sex forms pro portion of staminate flowers is very high than the pistillate/hermaphorodite flowers. Therefore, a high female to male ratio is desirable to increase fruit set and finally yield.

Control

1. Avoid over irrigation and excess application of nitrogenous fertilizers which lead perfuse vegetative growth and promote maleness
2. Application of growth substances/nutrients 2-3 times at 2-4 leaf stage which change the sequence of flowering
   a. Cucumber–Ethrel 250 ppm, Maleic hydrazide 100 ppm
   b. Bottlegurd-, Maleic hydrazide 100 ppm, 2-CEPA 100-200 ppm, boran 3 ppm
   c. Bitter gurd NAA100 ppm, 2-CEPA 100-200 ppm
   d. Ridge gurd- Ethrel 250 ppm
   e. Sponge gurd- Ethrel 250 ppm, Maleic hydrazide 100 ppm
   f. Watermelon- Maleic hydrazide 100 ppm
   g. Pumpkin- Maleic hydrazide 100 ppm

Unfruitfulness

Delayed ripening: It is found in muskmelon and water melon. The main causes of delay in ripening in these crops are high soil moisture, low temperature and fluctuation in temperature at ripening

Control

1. Adjust sowing time in such a way that fruit ripening coincides with hot and rain free period
2. Withhold irrigation at ripening stage

Delayed Ripening

It is found muskmelon and water melon. The main causes of delay in ripening in these crops are high soil moisture, low temperature and fluctuation in temperature at ripening.

Control

Adjust sowing time in such a way that fruit ripening coincides with hot and rain free periods. Withhold irrigation at ripening stage.

Salad Crops

Tip Burn

It is occurring due to prevalence of high temperature, light intensity for long duration, excess of nitrogen, calcium and boron deficiency. Tip burning is characterized by lateral margins of the inner leaves of mature heads

Control

Increase the dark period and relative humidity, Spray the crop with $CaCl_2$ @ 0.5 per cent

Spices

Black Pepper

Nutrient Deficiencies

Deficiency symptoms of macronutrients (Except Ca and S) were first appear on older leaves, while deficiency of micronutrients, symptoms first appear on younger leaves.

- ☆ **N**: Deficiency Symptoms appear as uniform yellowing followed by necrosis.
- ☆ **P**: Deficiency Symptoms appear as purple to bronze yellowing with ash coloured necrosis
- ☆ **Calcium**: Deficiency Symptoms appear as tiny brown necrotic spots on chlorotic area near margins which later enlarge and formed black necrotic areas

- ☆ **Mg**: Deficiency Symptoms appear as oval chlorotic area followed by black necrotic patches.
- ☆ **Sulphur**: Deficiency Symptoms appear as uniform yellowing with Brown necrotic spot.

Micronutrients

- ☆ **Iron:** Chlorosis was found by green bands along major veins
- ☆ **Mn**: Bronze yellow coloured of the intervienal area
- ☆ **Cu**: Bronze colour of entire lamina with necrotic tips and margins
- ☆ **Zn**: little leaf and resetting
- ☆ **B**: Leaves become large and thick and brittle with orange yellow mottles on upper surface and gray brown patches on lower surface.

Fennel

Sugary Secretion

It is a physiological disorder, due to which flowers secrete sugary substance; attracting fungus from outside. Due to this disorder, quality and production of crop is affected. This phenomenon is observed in those plots which have been highly manures and irrigated under heavy dewfall and cold condition. Sugary secretion spreads on flower parts, stem *etc.* Due to this mold and other fungus develop on plants causing black and gummy appearance. The sugary secretion attracts aphids too and these insects attack the plants. Combined effect of above gives false appearance of disease, but really sugary secretion is not a disease. For the control of this disorder, stop irrigation and manuring. Spraying of 0.03 per cent solution of Dimethoate or 0.05 per cent solution of Phosphamidone or any systemic pesticide should be done to control the aphids.

Chapter 32

Management

Integrated Disease Management

The intensive cultivation of genetically uniform and high yielding varieties of various crops coupled with excessive fertilization, close planting and frequent irrigation have contributed to aggravation of plant pathogens. Therefore, attention is focus on a system approached to keep the pathogens at a sub economic level with in population. A lot of information on this aspect is available which indicates that various approaches can be employed in judicious manner for integrated disease management

Management

1. Soil management
2. Seed management
3. Environmental management
4. Irrigation management
5. Crop management

Main Component of Integrated Disease Management

a) Prevention
b) Monitoring
c) Disease diagnosis
d) Knowledge of Acceptable disease threshold
e) Selection of management tools

Crop Management

a. Physical management
b. Cultural Management
c. Biological management
d. Chemical management

Crop rotation is single management practice, which is more effective in the elimination of soil borne pathogens. A micro culture of crops increases the population of microorganisms and permits them to alive on and within the root of those crops. In contrast, crop rotation help to prevent their build up, however pathogens of one crop are not likely to attack another unrelated crop. Adjustment of date of sowing is may be profitable practices to circumvent attack of disease by avoiding the peak period of attack. To put this concept into practice, the knowledge of conditions favouring appearance of disease and vulnerable age of the plant at which they are susceptible to attack is essential. The heat therapy is effective in case of internally seed borne diseases where mycelium remains in embryo in the dormant form. Hot water treatment eliminates mycelium of smut diseases. Mulching is used to reduce the over summering inoculum of the pathogen responsible for soil infestation. Plastic mulching is reduced the inoculums potential of *Pythium, Taphrina* and *Protomyces.* Excessive use of water and high doses of nitrogen in general prolong the period of crop which may be attacked by diseases. Nitrogenous fertilizer tends to increase susceptibility of rusts, mildews and blights. Genetic uniformity of varieties has been recognized by several workers to be crucial factors in facilitating disease epidemics caused by air borne pathogens like rust of pea. Genetic vulnerability can be avoided by deliberate creation of variation in the host population. The cultivation of diverse host is an effective method of plant disease management. Development and deployment of cultivars with durable resistance to plant diseases is a very economic and easy methods of plant disease management. Tools of molecular biology are now being used to facilitate the conventional disease resistance breeding programme and to shorten the duration required to develop a resistant cultivars in different crops. Molecular markers closely linked to resistant genes are used to identify the possible interogression of R-genes in susceptible cultivars upon hybridization. Selection of a particular trait in segregate population by the use of molecular markers is termed as Marker Assisted Selection (MAS).Use of DNA markers in plant disease resistance breeding programme have considerable increased in the country. There is a huge data on the molecular tagging of disease resistance gene irrespective of crop.

One of the common means of controlling plant pathogens in the field, green house and storage through use of chemicals that is toxic to the pathogens. Such chemicals are broad spectrum *i.e.* they are toxic to all or most of the pathogens. The efficacies of fungicides have been tested against pathogens in different crops. A disease prediction models is of fundamental importance for successful and efficient use of suitable chemicals for the managements of diseases. The feasibility of fungicide control has been restricted to economically protected situations and is not considered safe due to adverse effect on ecosystem. Hence, there is enthusiasm of searching mycoparasites to exploit tham as biological control agents. Antagonistic action of *Chaetomium globosum, Trichoderma* spp., *Gliocladium* spp., *Pseudomonas* spp., *Bacillus* spp., *Streptomyces* spp., *Penicillium* spp. and *Aspergillus* spp. has been found to be effective against many plant pathogens. Kalisena a bioformulation from strain of *Aspergillus niger* (AN-27) was found effective against many soil borne pathogens. Several phylloplane bacteria (belonging to Pseudomonas, Flavobacteria, Aeromonas) were used successfully as bio control agents for bacterial blight. Eco-friendly neem based products were effective as well as compatibility with chemical control. A few bacteriocin and HCN producing fluorescence pseudomonad were developed. The bacterisation of seed not only improved seed germinability but also reduced the root and seedling diseases of several crops.

Biopesticides

Agriculture is the main source of income of the rural masses in India and the disease accounts for more than 30 per cent losses every year. A number of microbes were found to be effective bio control agents for the disease

Origin of Biopesticides

As early as in 1920, it was noticed that pathogens are suppressed if some antibiotic producing microbes are added to the soil. However, the actual development began in 1963, when America organized an International Symposium on "Ecology of soil borne plant pathogens-Prelude to biological control"The proceedings was editated by Baker and Synder (1965) The first standing committee on biological control was establish within American phytopathological society in 1976. The Beltsville Agricultural Research Centre Symposium in USA was land mark where plant pathologist, entomologist weed scientists and nematologist were brought together for the first time. It is well known that the pesticides results in environmental contaminations as they remain over there for a longer periods. Therefore the government is attempting to increase the use of bio-pesticides using botanicals and microbial formulations. Farmers are also willing to adopt these pesticides with an intention to keep them safe and generate higher yields According to estimate only 5000 villagers comprising of 0.1 million farmers have reaped the knowledge and benefits occurring from the use of bio pesticides. Productions of organic Vegetables are coming in Indian scenario

Bacterial Antagonists

The widespread commercial success in the control of crown gall disease by an avirulent strain of *Agrobacterium radiobacter* K-84 has triggered the attention on the use of bacteria as biocontrol agent against plant pathogenic fungi. Antagonism is the phenomenon involved in the interaction. Antagonism can be described as killing, injury or inhibition of the growth of one microorganism by another. The possible modes of bacterial action in the inhibition of microorganisms are 1) direct penetration 2) antibiotic production 3) competetion for nutrient 4) stimulation of host defence and 5) antibiosis. Bacillus species are the promising bio control agents as they produce endospores that are tolerant to heat and desiccation. Seed bacterization with *Baccilus subtilis* has successfully controlled the disease of

Pseudomonas fluorescens

(Bacterial Bio agent -Multiplied in liquid culture)

Bacillus subtilis

(Bacterial Bio agent -Multiplied in liquid culture)

groundnut and onion. *Fusarium* wilt and *Pythium* sp. are controlled by the use of cell wall degrading lytic rhizosphere bacteria. It is also controlled by chitinolytic arthobacter species isolated from rhizosphere.Root diseases of crop plants are controlled by *Baccilus subtilis* with seed bacterization. Sneh *et al.,* evaluated over 700 bacteria and actinomycetes isolated from *Fusarium* suppressive soil for their ability to induce suppressiveness to *F.oxysporium f.* sp. *cucumerinum.* Among non-fluorescent bacteria only *Enterobacter cloacae* inhibited chlamydospore germination of *Fusarium.* They pointed out direct correlation between siderophore producting bacteria and inhibition of chlamydospore germination in soil. The *B. megaterium* B153-2-2 was a potential candidate for the management of microflora in rhizosphere. Bacterization of carnation roots with WCS417r controlled the *Fusarium* wilt by inducing phytoalexin accumulation.

Fungal Antagonist

Various fungi are involved in the bio control of soil borne diseases as they exhibit widespread mycoparasitism and antagonism towards pathogenic forms. Fungal bicontrol agents like *Trichoderma, Gliocladium, Penicillium* and *Aspergillus* are used for control of root diseases, several example are available for bio control of soil borne disease caused by *Fusarium, Sclerotium* and *Rhizoctonia* by species of *Trichoderma. Cucumber* wilt caused by *Fusarium cucumerinum* is controlled by *T. harzianum* and *P. flourescens.* Several other species of *Trichoderma* proved promising bio-agents against *F. moniliforme f.sp. licopersici* in tomato. Antibiosis, competition and mycoparasitism are the important phenomenon involved in the Trichoderma and plant pathogen interactions. *Gliocladium catenulatum* was found to be antagonistic against root pathogens by exerting similar antagonistic effect as that of *Trichoderma.*Tomato crown rot caused by *Fusariun oxysporum f. sp. radicislycopersici* was reduced by *Penicillium funiculosum* along with other antagonists in the field. Cabbage yellows disease incited by *F.oxysporum* is suppressed by *F.oxalicum* and *Trichoderma viride.* Pathogen of Water melon and Muskmelon wilt was suppressed by *Aspergillus niger. Aspergillus niger* was also found effective against brinjal and linseed wilt. Pre and post damping off was controlled by *T.viride, Streptomyces griseus and B. subtilis.* Antagonistic fungi are exposed to the fungicide applied to the soil.

Soil Solarization

Soil solarization is a method of hydrothermal disinfestations accomplished by covering moist soil with transparent plastic sheet during summer. The process is simple non hazardous and does not involve toxic materials; it is appropriate use of solar energy for soil solarization to kill the target organisms. This technology can effectively employed for controlling of soil borne diseases in tropical countries like India. Direct thermal effects are the major phenomenon implicated in the control of soil borne pathogen during solar heating. Moisture plays a crucial role in reducing the population of pathogens, thus soil irrigated before solarization gave higher lethal values. Soil solarization increased the antagonistic activity and inducing soil suppressiveness with increasing plant growth because major antagonistic microorganisms are agents of suppression to major and minor pathogens in soil through aggressive colonization of root and antibiotic production. Soil solarization

MASS MULTIPLICATION
OF
TRICHODERMA sp.
Trichoderma harzianum
(Multiplied in sterile coffee husk)

is a disinfestations method, has potential advantages. It is a non-chemical method, non-hazardous to the user and does not involve toxic substances, to the host plant or to other organisms. In this perspective, it is less expensive than other methods. Soil solarization technology can easily be transmitted by the farmers and can be applied in large areas manually and mechanically, used of this method has been reported to reduce the population of many soil borne pathogens including fungi, bacteria and nematodes as well as weeds. *Pythium aphanidermatum* was found favoured relatively high temperature of 35°C, which is approximately 5°C above that for intermediate *Pythium spp.* group. A clear and transparent polythene sheet used for soil solarization and plastic mulching may increase soil temperature upto 50°C, which killed the pathogens. Hence, soil solarization can helped for reduction of yield losses due to soil borne diseases.

Bio-fumigation

Bio-fumigation is used for the suppression of soil-borne pathogens and pests by the incorporation of *Brassica*, cabbage, cauliflower and radish leaves into the soil. The action of volatile compounds essentially isothiocyanates (ITC's) produced by the hydrolysis of glucosinolates which are found within *cruciferae*. Glucosinolates have little biological activity, however, products formed from their hydrolysis (isothiocynanates) are considered to be the most toxic. Suppression of some fungal pathogens by the ITC's liberated from *Brassica* plants has been shown to be superior that the synthetic fumigant metham sodium. Different crop of *cruciferece* are used as bio-fumigants because of their different concentrations and type of ITC's emission during their decomposition that are different in their toxicity against pathogenic fungi. The use of bio-fumigation with mustard induced high rate of rhizome germination along with reduced soft rot which has boosted the yield of ginger. These reductions were attributed due to production of volatile compounds from *Brassica* crops through a process known as bio-fumigation. It was observed that inoculums level of *P. aphanidermatum* reduced in bio-fumigated soil in comparison to untreated soil. Reduced level of damping-off disease was also found in greenhouse seedlings with bio-fumigation and enhanced crop growth such as stem height and stem diameter. Soil infested with *R. solani* have controlled, with the application of mustard seed meal and found that percentage of healthy plants was 20 per cent in comparison to 60 per cent pathogen-infested control treatment with seed meal mustard bio-fumigation. Organic amendments, particularly plant residues and animal manure, augment the biocidal activity of bio-fumigation through the production of biotoxic volatile compounds emanating from the decomposition of organic materials. The numbers of *P. ultimum* and *S. rolfsii* propagules were reduced by more than 95 per cent, when the exposed to volatile compounds from heated cabbage or compost amended soil for 14 days.

Botanical Pesticides

Green plants are known to provide a vast reservoir of potential drugs. More than 10 per cent of these plants have so for been examined in detail for their biological activity against human diseases and much less against plant diseases. In history poisonous plants are mention those have pest control properties like Indian

Rig Veda, The classic book of Hinduism, compiled during second millennium. Democritus tried plant extracts for controlling plant diseases as early as in 470 B.C. Decoction of tobacco and alder buds with lime and sulphur against mildew diseases of trees was recommended by Forsyth. Bradely used pepper dust to protect trees against blight to their blossom. Several plants are also known to possess antifungal properties against *F. udum*. Leaf extract of Neem (*Azadirachta indica*), Mehandi (*Lawsonia alba*), Onion (*Allium cepa*), Garlic (*Allium sativa*), Tulsi (*Ocimum sanctum*) and Eucalyptus (*E. globules*) has been found effective against wilt of pigeonpea caused by *F. udum*. The bioefficacy of phytoextracts, were tested *in vitro* against *F. udum* causing wilt of pigeon pea. The garlic bulbs extract was significantly superior to inhibit the growth of *F. udum*, which was followed by turmeric and neem leaves extract tested by poisoned food technique. Fresh leaf extract (5 per cent) of A. *indica*, *Dhatura fastousa* and *Tagetes erecta* which were found most effective against *F. udum*. I*n vitro* efficacy of bitter guard, turmeric, garlic and black papper have showed considerable diminution in the growth of *Fusarium udum*. Growth of *Fusarium udum* has been reduced by 15 per cent concentration of turmeric (89.2 per cent) followed by garlic (88.20 per cent) black papper (82.22 per cent). The bitter guard extract is found least effective against *F. udum*. Mandhare and Suryawanshi have screened 35 botanicals and found *Ocimum sanctum*, *Eucalyptus* spp. and *Nerium indicum* completely inhibited the growth of fungus on PDA as well as Potato Dextrose broth. The growth of *F. udum* was inhibited by 80, 75 and 71 per cent with extract of *Azadirachta indica*, flax line and *Hibiscus rosasinensis*, respectively. The extract of *Vitex negundo*, *Sorghum bicolor*, *Piper nigrum* and *Zea maysdis* were also effective and inhibited the growth of *F. udum* by 60, 60, 60 and 62 per cent, respectively. Neem Kernel extract (*A. indica*) was effective in suppressing mycelium growth of *F. udum* *in vitro* and controlling wilt disease *in vivo*. Garlic bulb crude extracts was reported to inhibit the spore germination and mycelia growth of different species of *Fusarium* causing wilt diseases in vegetable and horticultural. Leaf extracts of *Citrus media*, a root extract of *Asparagus adscendens*, rhizome extract of *Curcuma longa* rhizome extract of Zingiber *officinale* and a bulb extract of *Allium sativa* were found higher concentration of plant extract completely inhibited (100 per cent) the mycelial growth of *F. udum* both *in vitro* and *in vivo*. *Azadirachta indica* leaf extracts of 100 per cent concentration completely inhibited spore germination of *Fusarium* spp. while 100 per cent *Allium sativum* leaf extract partially inhibited spores of *F. equiseti* (20 per cent germination). *Ocimum sanctum* leaf extract 100 per cent inhibited the growth of fungus because it have antifungal properties, against the spores of *F. solani*, *F. oxysporum* and *F. equiseti* as compared to control, the inhibition was 90, 94 and 90 per cent respectively. Five medicinal plants (*Strychnos nux-vomica*, *Calotropis procera*, *Azadirachta indica*, *Ocimum sanctum* and *Allium sativum*) were evaluated on spore germination of different species of *Fusarium* and found complete inhibition by 100 per cent aqueous extract of *Azadirachta indica* while only 20 per cent germination of *F. euiseti* spores was recorded with treatment of 100 per cent aqueous extract of *Allium sativum*.

Equipment Used in Plant Disease Control

Pesticides used in different forms *viz.* spray, dust, fog, mist and smoke. On the basis of pest, used of chemical application, machine are essential to choose the right type of machine to performed effective management.

Plant protection machines can be classified as follows:

- ✰ Sprayer
- ✰ Dusters
- ✰ Sprayers cum dusters
- ✰ Seed dressing machines
- ✰ Fumigators
- ✰ Flame throwers
- ✰ Bird Scorers
- ✰ Mechanical seed eliminators

Sprayers

A sprayer breaks up the liquid in to droplets of effective size and distributed them uniformly over the surface or space to be protected. Another function is to regulate the amount of insecticides to avoid an excessive application that might prove harmful or wasteful. Nozzles are the most important. These should be selected on their ability to deliver the desired amount of spray and deposit it uniformly on the surface which has been treated. The common types of nozzles used in sprayers are flat fan, hollow cone and solid cone. The flat fan nozzles are essentially low pressure and low volume in character. A high pressure in this nozzle produces too many small drops which drift away. The flat fan nozzle delivers more spray towards the ends than in the centre or vice-versa after some time. It is used for especially weed control work. The hollow cone nozzle is made to liquid in swirl in a chamber, The film of liquid is escapes through a single hole and as it rotates. The performance of nozzle cone is affected by changing the swirl plates and altering chamber depth. These nozzles work with wide pressure range and are suitable for medium and high volume spraying. It does not give as uniform spray in the field. It is used for spraying in vegetable and field crops. The solid cone nozzles has the same mechanism as hollow cone nozzle except swirl plate bears a central hole It allow a faster application rate upto 5 g.p.m. at high pressure but it can work at low volume nozzle at low pressures. This type of nozzle produce large drops which vary very much in their sizes. It is suitable for spraying high trees from the ground. High concentration of the insecticide especially when formulation of water dispersible powders are used, affects the droplet size. The more concentration of powder in water, then less atomization and larger production of droplets. When spray liquid becomes sufficiently thick in concentration, it starts coming in the form of threads from nozzle instead of droplets and at time clogs the nozzle. The water dispersible powders are not considered suitable for low volume spraying.

Type of Sprayer

Hand Sprayer or Atomizers

These are two types one is single action and other is continuous action. Both develop 0.9-2.25Kg of air pressure and have 1/16 inch to 1/32 inch diameter of nozzle orifices and produce droplets 15-50 micron in size. They are used to apply sediment free spray materials. The single action type of sprayer acts only on the stroke of the pump and is provided with non adjustable nozzle. It is used for spot disinfection.

The continuous sprayers develops pressure after few strokes of the pump and gives continuous spray of uniform pattern, it have interchangeable and adjustable nozzles which are able to deliver fine to coarse droplet spray. These are suitable for spraying in small plants in kitchen gardens.

Compressed Air Sprayer

It has a tank to contain the liquid and air. Tank is fillied with liquid in ¾ capacities. In the liquid, the air pressure is created by means of a hand pump, which pressurizes the liquid from the top and forces it to go outsides in the form of spray. A working pressure of 22.5-31.5kg per square inch is created. It can discharge 45-91litre of spray liquid per hour. It is used for spraying field crops, vegetables and nursery

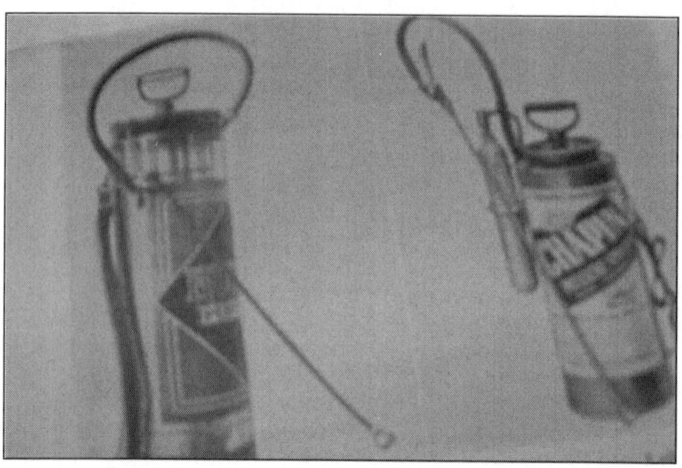

plants by carried on the back or shoulder of the operator. The essential parts are a pump with handle, pressure gauge, an air release, valve, liquid filling port, air jet agitator, back rest, delivery hose pipe, and spray lance with nozzles.

Battery Knapsack Sprayer

They are similar to compressed air sprayers except that air and water is charged into them through a central charge pump. The charge pump is hand or power driven. It need not to charge again and again and can be used conveniently.

The hand charge pump works like as ordinary pump for filling in air in the sprayer but when connected with a suction tube, it sucks liquid and pushes it into the sprayer tank till desired pressure is obtained. It can control about 6 sprayers. Such sprayer worked at a total pressure of 31.5-45kg./sq. inch. They are suitable for spraying all types of crops including tea and coffee but not grown up trees.

Knapsack Sprayer

This pump is operated constantly to get a continuous discharge of liquid. It is fit comfortable on the back of the operator. The lever handle provided at the bottom or on the top of the tank. A few stroke of the pump build up pressure in the air chamber located within the tank or outside. The handle is worked constantly so long sprayer is used. They build a pressures of 80-150 lbs with boom attachment it is possible to spray an area upto 2 acres/day with one sprayer.

Bucket Sprayer

It is designed to be used in a bucket containing spray liquid. It is provided with a pump which may be single or double acting. In latter case it discharges on both directions *i.e.* downwards and upwards strokes of handle Good sprayers can develop pressure upto 54kg/squire inch. It con used for small area in the garden or nursery beds.

Wheel Barrow Sprayer

It has a tank mounted on a single wheel or double wheel. The sprayer consist double acting hand pump fitted inside the tank. The capacity of the tank varies from 22.7- 68.1litre. It is operated by lever like handle and provides pressure inside the tank which can spray continuously without operating handles. The

Barrel Sprayer

pressure develops by this sprayer ranges from 90 to 135 kg/sq. inch. It is adequate for spraying orchards, medium sized fruit trees, nurseries, gardens and crops like cotton, sugarcane, potato in a day's 2-3 acres can be covered.

Wheelbarrow Sprayer

Slide Pump Sprayer

It is small hand operated sprayer comprising of telescopic plunger type pump. It have discharge pipe and suction hose with strainer. The sprayer is operated by back and forth motion of the plunger after dipping the suction hose in a bucket. It gives a continuous discharge of spray liquid and develops pressure upto 54kg/sq. inch (p.s.i.). It is suitable for ornamental pots, Kitchen garden and nursery plots.

Rocking Sprayer

It is provided with a pump and a pressure fitted on a wooden board and is operated by means of a handle connected with the pump. It may be double acting or single acting depending on the pump with one or two lances to spray. Rocking movement of handle helps in building up the pressure in the pressure chamber which forces the sucked in liquid into a spray through the nozzle. The part of sprayer is suction hose with strainer, a delivery hose with spraying lance. Before operating the sprayer, the suction hose is dipped in a liquid containing tank. It develops 90 to 112.5kg pressure and gives output of 68.1-90.8litre of liquid per hours with one nozzle. About 2 acres can be covered a day. It is suitable for medium size trees, crops and gardens.

Foot Sprayer

It is operated by foot with same principle as hand operated, in this sprayer constant pedaling is required after dipping the suction hose in a drum of bucket and

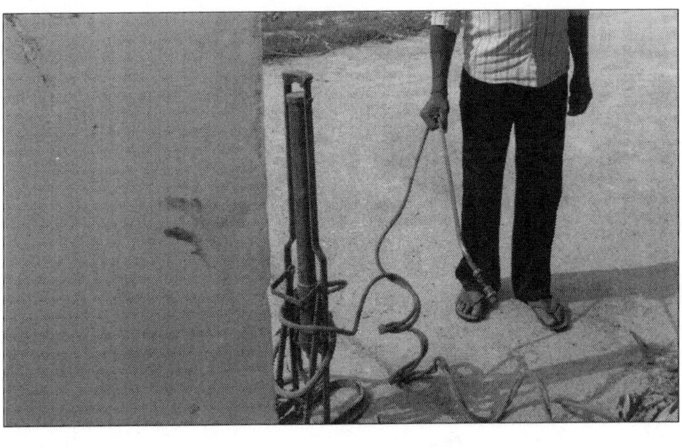

holding the sprayer with one or both hands. The pump is fitted on an iron stand and a pedal is attached with plunger rod and operates the sprayer by its upward and downward movements. Foot sprayer develops 112.5-135kg pressure p.s.i. it has an output of 113.5-136.2 litre/hour with one nozzle. About 2-3 acres of medium sized crop can be easily covered per day.

Power Sprayer

These are run by petrol or kerosene oil or electricity and are mainly three types-

1. Hydraulic sprayers
2. Mist sprayers
3. Fog or smoke generators

Hydraulic Sprayers

In hydraulic type power sprayers, the pressure created by the pump is forced out through the nozzle the spray liquid is generated in spray fluid. It is forced out through nozzles in small droplets. The droplets in this case are deposited on the surface by sheet force of the spray and are mostly 150-300 microns in size. The engine is generally 1 h.p to 20 h.p depending on the pump capacity and pressure range. Pump developing pressure upto 800 lbs p.s.i. or more can also be used. The satisfactorily at lower pressure range of 22.5-45 kg p.s.i. The essential parts of a hydraulic sprayer are engine, a pump, a tank with agitator, framework with or without wheels for mounting the engine, pressure regulator gauge, control valves, action and delivery hoses, strainer and nozzles. Hydraulic sprayer can be grouped into the following main types"

Multipurpose Farm Sprayer

These sprayers can work at reduced pressures at 30-50 lbs p.s.i. at high pressure at 90-360kg p.s.i. When spray required on weed or crop and fruit trees respectively. The tank has a capacity of 227-908 litre or more and is provided with a mechanical agitator. For spraying it is provided with booms or spray 5-10 acres/day.

General Use Sprayers

Under this category fall the small wheel barrow or skid type of sprayers with discharge capacity of 6.81 to 13.62 litre/min. at pressure upto 112.5kg. The power is furnished by on air cooled engine of 2/3 to 2 h.p and tank capacity varies from 12.5 to 18 gallons. They are provided with a mechanical or jet agitator, a pressure gauge, pressure regulator and a relief valve, spray gun, suction delivery hoses and

Engine Power Sprayer.

Field Crop Engine Sprayer.

handles for carrying it easily in the hand. A sprayer of this type can normally cover 3-5 acres/day.

High Pressure and High Volume Sprayers

These types of sprayers are used for spraying toll trees and dense crops. They discharge 36.32-272.4litre/minutes at pressure ranging upto 450kg. The tank size is 1362- 2724litre. The field of potato, tomato, cabbage and cauliflowers are used for pressure spraying. In a day 10-20 acres or more can be covered

Low Pressure and Low Volume Sprayers

For spraying pressure lower than 45kg p.s.i. low volume spraying have been desired to meet the requirement. They are used to spray crops with a smaller quantity of spray liquid at low cost. Their use is limited in as much as only sediment free type materials can be sprayed. They can be mounted on tractor a wheel mounted trailer. Such sprayers are provided with booms. One sprayer covers approximate 15-20 acre/days.

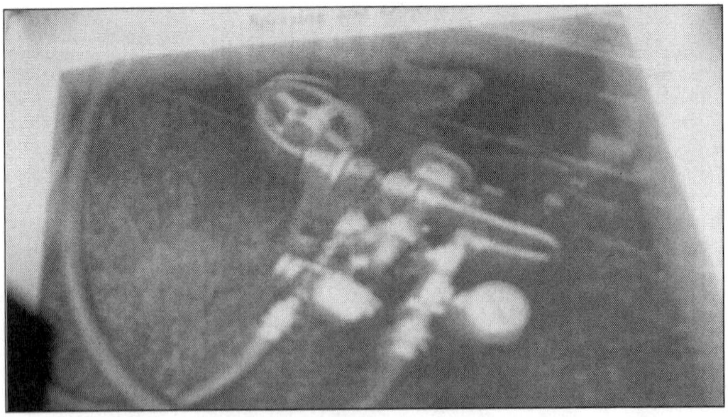

Low Pressure Sprayer

Self Propelled High Clearance Sprayer

They are used for spraying tall ground crops which cannot be sprayed by other conventional types of sprayers. Plants size upto 5-8 fit in height are easily sprayed by booms. The booms are adjusted accordingly to the height of the crop and are fed by generally a low pressure, low volume sprayer mounted on the self propelled carrying frame or tractor. Such sprayers have 3-4 wheels and power is applied to the rear wheel when tractor is not used for moving them. About 10-15 acres are covered/day.

Hydro Pneumatic Sprayer

There liquid is carried in a pressure tank and the spraying pressure is developed by means of an engine powered air compressor, thus avoiding the need to pass liquid to this pump. They are either skid type are mounted on wheels and work at maximum pressure of 33.75-45kg. In this sprayer discharge is denoted in cubic feet per minutes volume at the desired pressure equals 29.51 litre per minutes. It is not used for spraying in crops.

Mist Sprayers

They are essentially low volume sprayers developed to apply insecticides in a concentrated form and are known as air blast or drift sprayers. The dilution of spray materials is reduced by 20-80 per cent or more. Chemical saving is gave due to the production of smaller droplets 50-150 μ, which is a run off the from foliage. These are used for treating large areas of fruit trees, high and dense crops *etc.* The mist sprayer is fitted with a low pressure low volume pump which forces the spray materials under low pressure. It emits one gallons liquid per 10minutes and cover 150 to 250 acres in an hours.

Fog/Smoke Generators

They disperse spray material in extremely fine droplets (1-50 μ) in diameter which remain air borne for a long time. It can be successfully used in dense foliage or closed space where fog take time to settle and does not drift with the wind. It has the advantage of a uniform coverage of the object with chemicals.

The term high, low or medium volume spraying signifies the rate of application of the spray material over a given area. The rate of application is related to the crops being treated. The low volume sprayer is used for spraying citrus and mango and high volume is used for ground crops such as ground nut, paddy, potato, brinjal *etc.* The spray material is deposited over the surface of the leaves either in large drops or in small droplets. In high volume spraying the liquid will not spread in the form at a film then accumulation at the edge and finally run off on the ground. In medium spraying, the liquid tends to deposit on the leaf either in large drops or in small pools where large drops coalesce without runoff. When the spray is deposited in small drops discretely which do not form pools it becomes low volume spraying.

- ☆ High volume spraying- Over 272.4litres/acre
- ☆ Medium volume spraying- 90.8-272.4 litres/acre
- ☆ Low volume spraying- Upto 90.8litres/acre

The volumes are only guides. It is the coverage and spray deposit which are important for the control of insect pests or plant diseases. The bio efficiency of a deposit depends on:

a) The amount of active ingredient deposited
b) The distribution of active ingredient based on coverage of plant surface

The degree of coverage will depend on the nature of the organism to be controlled and the mode of action of the toxicant. Because highly mobile insect are need to cover thoroughly with chemicals. The main problems of high volume spraying is the obtaining and carting of huge quantities of water which is time consuming and add to the cost of spraying. Besides the output of work is less and the initial cost of high volume sprayer is more.

The principal advantage of low volume spraying lies in the saving to labour, active ingredient and time. In low volume spraying machines cast is comparatively less and output of work is high. Pesticides formulations are very important. The best type of formulations for use in low volume sprayers is the miscible liquid which could spray without sedimentation in the tank or blocking the nozzles for ting jets. They are three types:

1. Air blast spraying 227-2270litres/tank
2. Mist sprayers/blowers 45.4-90.8litres/acre
3. Drift sprayers 9.8-22.7 litres/acre

Advantages of Spraying

1. Greater range of formulation is available
2. Better utilization of chemicals
3. Rate of application per unit area can be increase or decrease with droplets size
4. Uniform deposition and better adherence
5. Cost of spraying is less

Disadvantages of Spraying

1. Less output of work per unit area
2. Equipment is generally heavier
3. Preparation of pesticide formulation is difficult
4. Spray formulation is requires carefully storage
5. Application cost is high

Dusters

Pesticides may be applied in dust formulations with specially designed machines called dusters, which are mainly hand driven and sometimes power driven. All dusters essentially consist of a container which is termed as a hopper of a system of keeping the dust particles agitated and in motion and to be fed into a current of air so that the particles are ultimately discharged in the form of fine dust.

Type of Dusters

1. Hand dusters
    a. Package or container dusters
    b. Hand pumps
    c. Bellown type dusters
    d. Rotary hand dusters
2. Power dusters
    a. Tracion duster
    b. Engine operated power duster

Container Dusters

These are primitive. The dust is kept inside a plastic rubber of leather containers. By squeezing or through a crude valve system dust is discharged into a small stream. These are useful for kitchen garden and for domestic use.

Hand Pumps

They are small and operated by a plunger pump. The container is normally straight tube with nozzle. Dust is thrown out through the outlet by the compression stroke of plunger pump. These dusters have very limited use in kitchen gardens, houses or on animals. Plunger dusters are simple in design and consist of an air pump.

Bellow Type

They are operated by the expansion and contraction of a pair of bellows during which process dust in sucked in and then thrown out into the delivery system. A

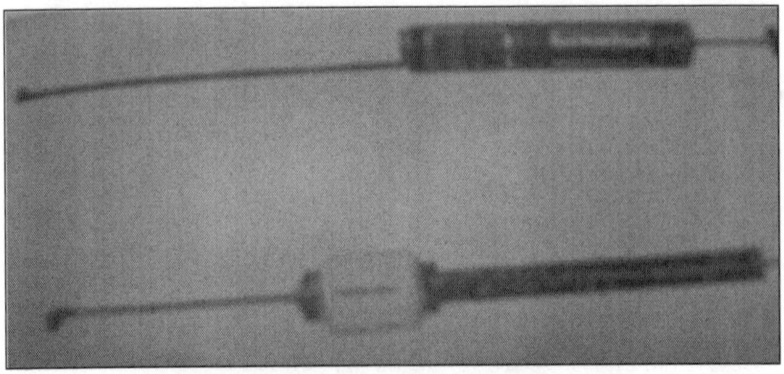

Plunger Type Hand Duster.

more convenient type of bellow duster is designed in the knapsack fashion. The hopper is of longer capacity.

Rotary Hand Dusters

It is also known as Crank type duster which is extensible used for dust application over large area of crops. These duster operated by the current at air produced by a fan which is driven by the rotary motion of a handle through a reduction gear. The fans draws dust from the dust chamber and discharge it into the delivery pipe from where it is blown out through one row or two row nozzles. Normally 0.5 to 150 gm/minutes may be discharged at 35 rpm. 2 to 3 acres can be covered with rotary duster in a day.

Fan Type Hand Duster.

Power Dusters

The basic components of a power duster are essentially the same as in a hand duster. The power for operation is generated from the wheels, engine or vehicles on which they are mounted. The power dusting machine comprises a hopper containing one or more agitators, a regulator for discharging of dust, a fan or blower producing air blast for drawing the dust from hopper and blowing its outside. The size of hopper varies with capacity to holding 60 to 200 pounds of dust depending on the density.

Power Duster.

Wheel-barrow Dusters

These types of dusters are provided one or two wheels and drawn by hands or animals. Such dusters have greater hopper capacity ranging from 9-22.5kg and can be used for dusting row crops in larger area, where crank or power dusters are not economical.

Tractor Mounted Dusters

The tractor mounted dusters deriving power from the tractor are provided with power from the tractor are provided with power take off sleeve coupling or a belt and pulley for mounting directly on the power take- off shaft. No separate power source is needed. The capacity of the hopper varies from 20-45kg.

Engine Operated Dusters

Normally 1-3 h.p. air cooled engines are used and the hopper capacity is usually 10-20 kg. Power duster may have a single outlet or may be fitted a series of 4-8 outlets with flexible pipes and fitted with a boom with as many as 18 nozzles. Dust can be discharged at the rate of 1-9 kg/min. A power duster with one out let may cover 12ha/hr. dust flame may be regulated from 0-1.5 kg/min by adjustment at multi-hole discs. Large sized dusters with hopper capacity of 50-100 kg are operated by engines upto 25 hp with the introduction of low volume spraying. The concentrate spraying materials of dusting is no longer palpated to the extent as it was before.

Knapsack Type Spray Dusters

This type of duster is operated by 1 h.p. to 2 h.p. engines and can be carried on the back of the operator. The hopper and the delivery hose pipe in such types of machines are small and feeding capacity is 4.5-5.4kg of dust and dust discharge is about 1 to 4 pound per minutes. Such type of duster is operated by one man. A spray dust can be done 10 acres per day.

Dusting in Potato Field.

Advantage of Duster

1. Less deterioration of dust formulation
2. Dusting can be done in high wind velocity
3. Dusting equipment is simpler
4. Less cost of treatment

5. High work can be done per unit time

Disadvantage of Dusters

1. Higher cost of active ingredient
2. Un even deposition and poor adhesion
3. Range of particle size is fixed
4. High transportation cost due to bulk of dust
5. Dusting can be done only morning hours

Seed Deressors

Many type of seed dressers are available and they vary in capacity

Type of Seed Dressers

1. Rotary seed dresser
2. Cascade or gravity seed dresser
3. Slurry seed dresser
4. Grain treating machine
5. Sack impregnating machine

Rotary Seed Dresser

It consists of a drum, fitted on a stand which can be rotated with the help of a handle provided at one end. The drum has opening to introduce seed and chemical, which take out seed after treatment. The drum should be rotated slowly 30-40 times during the process seeds will be thoroughly coated with the chemical.

Gravity Seed Dresser

The machine has a steel drum fitted with cones and baffles which are perforated. A lid is provided at the top through which the seed and chemicals are introduced. Seeds while coming downwards slowly due to gravity get mixed with the chemical. This is used for chemicals which are comparatively volatile in nature. It take more time but does not involve any manual labour.

Slurry Seed Treatments

These machines are driven by 3-4 hp electric motor or oil engine measured quantities of seed and chemical (slurry) are automatically fed into the treatment chamber in a synchronized manner. The slurry cap introduces a given amount with each seed lot into the mixing chamber. The seed is thoroughly mixed in the chamber to prevent sedimentation. Approximately 30 quintals of seed may be treated per hours in this machine.

Grain Treating Machine

This is designed by PAU Ludhiana. It has a hopper, grain metering device and a spraying unit. The spraying unit has a low volume nozzle; the spraying liquid is discharged at the rate of 290 ml/min. at a pressure of 2.1 Kg/cm^2. As grains which are fed on the top fall slowly due to gravity. They are treated with the chemicals these machines can treat upto 100 quintals of grains in 2 hours.

For Pulse

The pulse grains from the hoppers are moved through corrugated rollers into collecting container through a trough, while the grains are treated with chemicals, which is applied by a hand compression sprayer. About 100Kg of grain are fed in a period of 8-10 minutes.

Sack Impregnating Machine

This method has been developed by the CFTRI, India. Here the gunny bugs meant for storage of grains are treated. The sacks move through an endless they are impregnated with chemical which is applied from compressor in 10 liter tank maintained at constant pressure. The liquid is applied through low volume nozzles only one side is treated at a time approximately 3000 sacks can be treated/day.

Soil Injectors

Types

1. D.D.gun
2. Tractor mounted applicators
3. Granule applicator
4. Aerosol projectors

D.D. Gun

It consists of pump, container for the liquid to be injected and Holland injection needle. The nozzles can be driven up to depth of 25-30 cm. The pump then pushed down to force liquid into soil containers have capacity of 2.2 to 3.5 liters.

It is used for kitchen garden or experimental purposes.

Tractor Mounted Applicators

Field applicators consist of container from which large fines come out. They are drowning like harrow. The liquid is fed into the fines as the fines move through soil.

Granule Applicator

The essential component are the hopper in which the granules are kept, a long flexible discharged tube with nozzle at the distal end and a finger controlled mechanism for the regulation of consistent flow of granules. They are usually made of plashes or light and have a granules capacity of 1.0 to 1.3 Kg.

Dust Applicator, Cymag Foot Pump

It is essentially a plunger type at duster with a dust chamber of 450g capacity, pump, and discharge pipe and delivery tube. The air blast agitates powder and forces the same to be discharged through delivery tube.

Aerosols Projector

These are useful in large green houses enclosed spaces, go downs and store houses. Aerosol formulation is poured into the pressure chamber. Gas is liberated which exert pressure on the aerosol liquid on release of pressure by trigger. The liquid is forced through the nozzles overhear. The liquid are reaching outside because of low boiling point vaporizes immediately with the formulation of fine mist.

Methods of Disease Management

Seed Treatments

Mechanical Separation

The seed lots are cleaned mechanically by separating the soil clods, plant debrish, weeds, discoloured, deformed and shriveled seeds showing visual symptoms of the disease by hand picking under stereobinocular microscope.The vegetative propagules are cleaned by excising the infected portions.

Hot Water Treatments

This technique is used for controlling various seed-borne pathogens. In this treatment, the seed with or without pre-soaking, is deep in water at certain temperature generally close to the maximal temperature critical for both pathogen and host for a definit periods of time. Wet heat appears to be an effective agent for directly killing the pathogen.

Dry Heat Treatment

Dry heat treatment has been used by some workers occasionally. Different combinations of temperatures have been tried which given promising results against certain pathogens.

Solar Heat Treatment

This treatment is applicable in countries with hot climate and plenty of sunshine. Grain is pre-soaked in water for some hours under shade or in room and then dried on ground in the layers under bright sunlight, usually at noon.

Vapour Heat Treatment

The dry seed are treated in vapour condition generally by pressure flow. It has also been used to treat the vegetative propagules for eradication of virus infection.

Chemical Treatment

Chemicals are used for controlling seed borne pathogens. Various fungicides are available which can be used as seed dressing or dip for the vegetative propagules. The chemicals also give protection against soil borne fungi. Systemic fungicides are widely used at present. The metalazyl seed treatment has been effective against the powdery mildew. Oxathiins Plantvax and vitavax are used against smuts. Benomyl compound are broad spectrum and are effectives against many seed borne fungi. Fungicides can applied with following methods:

Steep Treatment

The seed is soaked in fungicidal liquid solution or suspension for a definite period of time. Fungicides solution is then drained and seed is dried.

Sprinkling Treatment

The seed is sprinkled with a fungicidal liquid, solution or suspension, left and damp with this treatment for a definite period of time and then dried.

Dust Treatment

The seed is mixed with a fungicides dust in a mixing device until the seed is well coated usually @ 2 g/Kg seed.

Slurry Treatment

The seed is mixed with a dust fungicide in a special treater in which small calibrater amount of liquid 5-10 ml/Kg seed is added, thus forming a soup like slurry to ensure coating without undue wetting.

Wet Treatment

The seed is mixed with relatively small amount of concentrated liquid about 100-300 ml/kg seed. There is no recovery of liquid after treatment but drying is necessary.

Fumigation

The seed is treated in an airtight container or chamber for a definite period of time using volatile materials. A fungicides, nematicides or insecticides.

Pelleting

Application of sticker is necessary prior to mixing the seed with a dust which may contain pesticides or other materials. The seed is first coated with methocel. A cellulose acetate sticker is in dilute solution and then agitated with the dust in such a way that each seed becomes a pellets containing an outer zone of protectant.

Sprays and Dusts

Chemicals applied as spray or dusts on the plants are usually aimed for control to pathogens. Most fungicides and bactericides are protectants and present on the surface of plant in advance of pathogen, that can prevent to entry of pathogens. Because their present on the surface of plant does not allow pathogens spores to germinate and they may kill spores. Some fungicides have direct effect on pathogens which have already invaded the leaves, stem, root and fruit in this situation fungicides act as eradicant and killing the pathogens inside the host or they may suppress the sporulations and multiplication. Some fungicides are partially systemic and some are clearly systemic, partially systemic fungicides are absorbed by aparts of plant leaf tissues and translocated internally into the whole leaf area example dodine. Systemic fungicides translocated internally throughout the host plant and kill the pathogens *e.g.* benzimidazoles, benomyl, thiobendazole, oxanthiins, carboxin and oxycarboxin.

Spray of fungicides is more effectives than dusting of fungicides. Dusting is preferable during rainy weather because of adhere better to wet plant tissues.

Drenching

The oldest and best method of chemical treatments is the drench especially formaldehyde drench. 4.54 litre of formaldehyde (37-40 per cent) is added to 227 litre of water or incase of heavy disease infectations the strength may be increased to 4.54 litre of formaldehyde to 113 litre of water. The soil must be drenched with this solution to a depth of 3-6 inches. This require from 115-130 litre of solution per square feet of surface. The soil surface should be covered with polythin sheet for a 48 to 72 hrs after treatment. The soil should be forked or spaded before treatment. After treatment the soil must be aired until all traces of formaldehyde odor have disappeared. It required 10-15 days. The present of small amount of chemical in the soil will kill the young seedlings. This method is effective for soil borne disease management.

Fumigation

Fumigation is especially valuable as a method of controlling disease, nematode and pest in soil, greenhouse, warehouse, mills *etc.* where spraying is not possible and controlling insect, nematodes and diseases. The chemicals used for this purpose are usually volatile and on coming in contact with soil moisture release gases which diffuse in the soil and kill the larvae of the nematode and insect and spore of fungi. Application of these high volatile toxic substances is recommended few days advance sowing and planting of crops. Fumigating equipment is required for injecting the chemicals in soil. The diffusion of chemical in soil is restricted at regular distance 12-15 inches. The depth of application is kept at 6-9 inches. The effectiveness is depending on soil texture depth, soil temperature, moisture and pulverization.

Paste and Paint

Wounds created during hoing, pruning and training of trees often serve as openings for entry of pathogens. The protection of these wounds required protective layers of past and paint. These pastes are prepared with fungicidal chemicals such as copper sulphate and lime, copper corbonate, red lead and antibiotics *etc.* in suitable carriers. The past heel cut surface of wound for long enough and protect the entry of pathogens.

Basic Precaution in Pesticides Usages

A. Purchase

- ☆ Purchase only just required quantity *e.g.* 100,250,500 or 1000g/l for single application in specified area.
- ☆ Do not purchase leaking containers, loose, unsealed or torn bags
- ☆ Do not purchase pesticides without proper/approved labels

B. Storage

- ☆ Avoid storage of pesticides in house premises

Identification of Pesticides Hazardous and Symptoms of Poisoning

Sl.No.	*Colour of Toxicity Triangle in Pesticides*	*Classification as per Pesticides Rule,1971*	*WHO Classification by Hazard*	*Symptoms of Poisoning*
1	Yellow	Highly toxic	Class-II, Moderate hazardous	Nausea, vomiting, respiratory failure, coma, weakness, aniety and death
2	Red	Extremely toxic	Class-1b, highly hazardous	Constriction of pupls, salivation, profuse sweating,vomiting diarrhea, epigastric, tightness in cheas
3	Green	Slightly toxic	Table-5(unlikely to present acute hazar in normal use)	Headache, palpitation, vomiting, irritation of nose, throat, eyes and skin
4	Blue	Moderate toxic	Class-II slightly hazardous	Nausea, vomiting, irritation

- ☆ Keep only in original container with intact seal
- ☆ Do not transfer pesticides in other containers
- ☆ Naver keep them together with food, feed/fodder
- ☆ Keep away from reach of children and livestock
- ☆ Do not expose to sun-light or rain water
- ☆ Do not store weedicides along with other pesticides

C. Handling

- ☆ Never carry/transport pesticides along with food materials
- ☆ Avoid carrying bulk-pesticides on head, shoulders or on the back

D. Precautions for Preparation of Spray Solution

- ☆ Use clean water
- ☆ Always protect your nose, eyes, mouth, ears and hands
- ☆ Use hand gloves, face mask and cover your hands with cap
- ☆ Use polythin bags as hand gloves, hand kerchiefs or piece of clean cloth as mask and a cap or towel to cover the head
- ☆ Read the level on the container before preparing spray solution
- ☆ Prepare solution as per requirement
- ☆ Do not mix granules with water
- ☆ Concentrated pesticides do not fall on hands *etc.* while opening sealed containers.
- ☆ Do not smell the pesticides
- ☆ Do not spilling of pesticides solution while filling the spray solution tank
- ☆ Do not eat, drink smoke or chew while preparing solution
- ☆ The operator should protect his bare feet and hands with polythin bags

E. Equipment

- ☆ Select right kind of equipment
- ☆ Do not use leaking, defective equipment
- ☆ Select right kind of nozzle
- ☆ Do not blow/clean clogged- nozzle with mouth
- ☆ Use old tooth brush tied with the sprayer and clean with water
- ☆ Do not use same sprayer for weedicides and pesticides

F. Precaution for Applying Pesticides

- ☆ Apply only recommended dose of pesticides
- ☆ Do not apply on hot sunny day or strong windy condition
- ☆ Do not apply just before the rains and also after the rains

- ✰ Do not apply against the wind direction
- ✰ Emulsifiable concentrate formulations should not be used for spraying with battery operated ULV sprayer
- ✰ Wash the sprayer and buckets *etc.* with soap water after spraying
- ✰ Containers, buckets *etc.* used for mixing pesticides should not be used for domestic purposes
- ✰ Avoid entry of animals and workers in the fields immediately after the spraying

G. Disposal

- ✰ Left over spray solution should not be drained in ponds or water lines *etc.* Throw it in barren isolated area if possible
- ✰ The used/empty containers should be crushed with a stone/stick and buried deep into soil away from water source
- ✰ Never re-use empty pesticides containers for any purpose

H. Warnings

- ✰ Consult expert of the field of plant protection before use of any pesticides regarding doses and time of application
- ✰ Do not used pesticides in wrong way
- ✰ Select pesticides wise fully
- ✰ Use pesticides judiciously on need based manner
- ✰ Do not use green leaves as food after spray of pesticides

Method of Pesticides and Fungicides Calculation

General Formula

$$F = \frac{Sc}{ai}$$

where,

F = is the quantity of formulation of fungicides or pesticides is required

S = is quantity of spray fluid required

C = is the per cent concentration of the fungicides or pesticides in the spray fluid

a.i = is the per cent concentration of active ingredient in the formulated fungicides or pesticides

Example No. 1

Calculate the amount of Bavistin 50WP required to prepare 10 liter of 0.1 per cent spray solution

F = ?

S = 10L.

C = 0.1 per cent

ai = 50

$$F = \frac{Sc}{ai}$$

$$F = \frac{10 \times 0.1}{50}$$

= 0.02Kg or 20gm

Example No. 2

How much spray fluid of 0.1 per cent concentration can be prepared from 125g of Bavistin 50WP

F= 125g

S = ?

C = 0.1 per cent

ai = 50

$$F = \frac{Sc}{ai}$$

$$125 = \frac{S \times 0.1}{50}$$

$$S = \frac{125 \times 50}{0}.1$$

=62500ml or 62.5litre

Example No. 3

Thiodan 35 E.C. measuring 300 ml is added in 150 litres of water. Determine the per cent concentration of the pesticides in the spray fluid

F= 300ml

S = 150L

C = ?

a i = 35

$$F = \frac{Sc}{ai}$$

$$300 = \frac{150000 \times c}{5035}$$

$$C = \frac{300 \times 35}{150000}$$

= 0.07 per cent

Example No. 4

Determine the per cent of a i in Roger when 400 ml of the formulation is diluted with 300 litre of water to get a concentration of 0.04 per cent for spraying against pea leaf minor

F= 400 ml

S = 300L.

C = 0.04 per cent

a i = ?

$$F = \frac{Sc}{ai}$$

$$2/5 = \frac{300 \times 0.04}{ai}$$

$$ai = \frac{5 \times 300 \times 0.04}{2}$$

= 30 per cent

Example No. 5

How much quantity of Baslin 20 E C is required for spraying 0.25 per cent spray fluid at the rate of 100 litre/acre for controlling the dimond black moth on cauliflower over an area of 1/10 th of an acre

F= ?

S = 100 x 1/10 = 10 litre

C = 0.025 per cent

ai = 20

$$F = \frac{Sc}{ai}$$

$$F = \frac{10 \times 0.025}{20}$$

$$= \frac{1}{80}$$

= 12.5 liter

Insecticides Registered Under/Section 9 (3) of the Insecticide Act, 1968 as on 17/06/2011

Sl.No.	Name of the Pesticide
1	2,4-Dichlorophenoxy Acetic Acid
2	Acephate
3	Acetamiprid
4	Alachlor
5	Allethrin
6	Alphacypermethrin
7	Alphanaphthyl Acetic Acid
8	Aluminium Phosphide
9	Anilophos
10	Atrazine
11	Aureofungin
12	Azadirachtin (Neem Products)
13	Azoxystrobin
14	Bacillus sphaericus
15	Bacillus thuringiensis var. israelensis
16	Bacillus thuringiensis var. kurstaki
17	Bacillus thuringiensis var. galleriae
18	Barium Carbonate
19	Beauveria bassiana
20	Bendiocarb
21	Benfuracarb
22	Benomyl
23	Bensulfuron Methyl
24	Beta Cyfluthrin
25	Bifenazate
26	Bifenthrin
27	Bispyribac Sodium
28	Bitertanol
29	Bromadiolone
30	Buprofezin
31	Butachlor
32	Captan
33	Carbaryl
34	Carbendazim
35	Carbofuran
36	Carbosulfan
37	Carboxin
38	Carfentazone Ethyl

Sl.No.	*Name of the Pesticide*
39	Carpropamid
40	Cartap Hydrochloride
41	Chlorantraniliprole
42	Chlorfenapyr
43	Chlorimuron ethyl
44	Chlormequat Chloride (CCC)
45	Chlorothalonil
46	Chlorpropham
47	Chlorpyriphos
48	Chlorpyriphos Methyl
49	Cinmethylene
50	Clodinafop-propargyl
51	Clomazone
52	Chlothianidin
53	Copper Hydroxide
54	Copper Oxychloride
55	Copper Sulphate
56	Coumachlor
57	Coumatetralyl
58	Cuprous Oxide
59	Cyfluthrin
60	Cyhalofop-butyl
61	Cymoxanil
62	Cypermethrin
63	Cyphenothrin
64	Dazomet
65	Deltamethrin (Decamethrin)
66	Diazinon
67	Dichloro Diphenyl Trichloroethane (DDT)
68	Dichloropropene and Dichloropropane mixure (DD mixure)
69	Diclorvos (DDVP)
70	Diclofop-Methyl
71	Dicofol
72	Difenocenazole
73	Difenthiuron
74	Diflubenzuron
75	Dimethoate
76	Dimethomorph
77	Dinocap
78	Dithianon

Sl.No.	*Name of the Pesticide*
79	Diuron
80	Dodine
81	D-trans Allethrin
82	Edifenphos
83	Emamectin Benzoate
84	Endosulfan*
85	Ethephon
86	Ethiprole
87	Ethion
88	Ethofenprox (Etofenprox)
89	Ethoxysulfuron
90	Ethylene Dichloride and Carbon Tetrachloride mixture (EDCT Mixture 3:1)
91	Famoxadone
92	Fenamidone
93	Fenarimol
94	Fenazaquin
95	Fenitrothion
96	Fenobucarb (BPMC)
97	Fenoxaprop-p-Ethyl
98	Fenpropathrin
99	Fenpyroximate
100	Fenthion
101	Fenvalerate
102	Fipronil
103	Fluazifop-p-butyl
104	Flubendiamide
105	Fluchloralin
106	Flufenacet
107	Flufenoxuron
108	Flufenzine
109	Flusilazole
110	Fluvalinate
111	Forchlorfenuron
112	Fosetyl-Al
113	Gibberellic Acid
114	Glufosinate Ammonium
115	Glyphosate
116	Hexaconazole
117	Hexazinone
118	Hexythiazox

Sl.No.	Name of the Pesticide
119	Hydrogen Cyanamide
120	Imazamox
121	Imazethapyr
122	Imidacloprid
123	Imiprothrin
124	Indoxacarb
125	Iprobenfos (Kitazin)
126	Iprodione
127	Iprovalicarb
128	Isoprothiolane
129	Isoproturon
130	Kasugamycin
131	Kresoxim Methyl
132	Lambdacyhalothrin
133	Lime Sulphur
134	Linuron
135	Lufenuron
136	Magnesium Phosphide Plates
137	Malathion
138	Mancozeb
139	Mepiquate Chloride
140	Mesosulfuron Methyl + Iodosulfuron Methyl Sodium
141	Metaflumizone
142	Metalaxyl
143	Metalaxyl-M
144	Metaldehyde
145	Methabenzthiazuron
146	Methomyl
147	Methoxy Ethyl Mercury Chloride (MEMC)
148	Methyl Bromide
149	Methyl Chlorophenoxy Acetic Acid (MCPA)
150	Methyl Parathion
151	Metiram
152	Metolachlor
153	Metribuzin
154	Metsulfuron Methyl
155	Milbemectin
156	Monocrotophos
157	Myclobutanil
158	Novaluron

Sl.No.	Name of the Pesticide
159	Nuclear polyhyderosis virus of Helicoverpa armigera
160	Nuclear polyhyderosis virus of Spodoptera litura
161	Oxadiargyl
162	Oxadiazon
163	Oxycarboxin
164	Oxydemeton-Methyl
165	Oxyfluorfen
166	Paclobutrazol
167	Paraquat dichloride
168	Penconazole
169	Pencycuron
170	Pendimethalin
171	Permethrin
172	Phenthoate
173	Phorate
174	Phosalone
175	Phosphamidon
176	Prallethrin
177	Pretilachlor
178	Primiphos-methyl
179	Profenophos
180	Propanil
181	Propaquizafop
182	Propergite
183	Propetamphos
184	Propiconazole
185	Propineb
186	Propoxur
187	Pyrachlostrobin
188	Pyrazosulfuron ethyl
189	Pyrethrins (pyrethrum)
190	Pyridalyl
191	Pyriproxyfen
192	Pyrithiobac sodium
193	Quinalphos
194	Quizalofop ethyl
195	Quizalofop-P-tefuryl
196	S-bioallethrin
197	Sirmate
198	Sodium Cyanide

Sl.No.	*Name of the Pesticide*
199	Spinosad
200	Spiromesifen
201	Streptomycin + Tetracycline
202	sulfosulfuron
203	sulphur
204	Tebuconazole
205	Temephos
206	Thiacloprid
207	Thifluzamide
208	Thiobencarb (Benthiocarb)
209	Thiodicarb
210	Thiomethoxain
211	Thiometon
212	Thiophanate-Methyl
213	Thiram
214	Transfluthrin
215	Triacontanol
216	Triadimefon
217	Triallate
218	Triazophos
219	Tricholorofon
220	Trichoderma harzianum
221	Trichoderma Viride
222	Tricyclazole
223	Tridemorph
224	Trifloxistrobin
225	Trifluralin
226	Validamycin
227	Verticillium lecanii
228	Zinc Phosphide
229	Zineb
230	Ziram

Endosulfan*:- Endosulfan has been banned by the supreme Court of india w.e.f. 13-05-2011 for production, use and sale all over India till further orders vide ad-Interim order in the Writ Petition (Civil) No. 213 of 2011.

Pesticides and their formulation registered under the Insecticides Act

Insecticides registered under the Insecticides Act for use in Household/Public Health/Rodent *etc.*

Compendium of Registered pesticides, source of supply and list of manufacturers under Section 9(3) of the Insecticides Act, 1968

Note: The list is subject to change from time to time.

List of Pesticides which are Banned, Refused Registration and Restricted in Use (As on 20th October 2015)

I. PESTICIDES/FORMULATIONS BANNED IN INDIA

A.	**Pesticides Banned for manufacture, import and use.**	
	1.	Aldicarb (vide S.O. 682 (E) dated 17th July 2001)
	2.	Aldrin
	3.	Benzene Hexachloride
	4.	Calcium Cyanide
	5.	Chlorbenzilate (vide S.O. 682 (E) dated 17th July 2001)
	6.	Chlordane
	7.	Chlorofenvinphos
	8.	Copper Acetoarsenite
	9.	Dibromochloropropane (DBCP) (vide S.O. 569 (E) dated 25th July 1989)
	10.	Dieldrin (vide S.O. 682 (E) dated 17th July 2001)
	11.	Endrin
	12.	Ethyl Mercury Chloride
	13.	Ethyl Parathion
	14.	Ethylene Dibromide (EDB) (vide S.O. 682 (E) dated 17th July 2001)
	15.	Heptachlor
	16.	Lindane (Gamma-HCH)
	17.	Maleic Hydrazide (vide S.O. 682 (E) dated 17th July 2001)
	18.	Menazon
	19.	Metoxuron
	20.	Nitrofen
	21.	Paraquat Dimethyl Sulphate
	22.	Pentachloro Nitrobenzene (PCNB) (vide S.O. 569 (E) dated 25th July 1989)
	23.	Pentachlorophenol
	24.	Phenyl Mercury Acetate
	25.	Sodium Methane Arsonate
	26.	Tetradifon
	27.	Toxaphene(Camphechlor) (vide S.O. 569 (E) dated 25th July 1989)
	28.	Trichloro acetic acid (TCA) (vide S.O. 682 (E) dated 17th July 2001)
B.	**Pesticide formulations banned for import, manufacture and use**	
	1.	Carbofuron 50 per cent SP (vide S.O. 678 (E) dated 17th July 2001)
	2.	Methomyl 12.5 per cent L
	3.	Methomyl 24 per cent formulation
	4.	Phosphamidon 85 per cent SL

C.	**Pesticide/Pesticide formulations banned for use but continued to manufacture for export**	
	1.	Captafol 80 per cent Powder (vide S.O. 679 (E) dated 17th July 2001)
	2.	Nicotin Sulfate
D.	**Pesticides Withdrawn** **(Withdrawal may become inoperative as soon as required complete data as per the guidelines is generated and submitted by the Pesticides Industry to the Government and accepted by the Registration Committee. (S.O 915(E) dated 15th Jun,2006)**	
	1.	Dalapon
	2.	Ferbam
	3.	Formothion
	4.	Nickel Chloride
	5.	Paradichlorobenzene (PDCB)
	6.	Simazine
	7.	Sirmate (S.O. 2485 (E) dated 24th September 2014)
	8.	Warfarin (vide S.O. 915 (E) dated 15th June 2006)

II. PESTICIDES REFUSED REGISTRATION

Sl.No.	*Name of Pesticides*
1	2,4, 5-T
2	Ammonium Sulphamate
3	Azinphos Ethyl
4	Azinphos Methyl
5	Binapacryl
6	Calcium Arsenate
7	Carbophenothion
8	Chinomethionate (Morestan)
9	Dicrotophos
10	EPN
11	Fentin Acetate
12	Fentin Hydroxide
13	Lead Arsenate
14	Leptophos (Phosvel)
15	Mephosfolan
16	Mevinphos (Phosdrin)
17	Thiodemeton/Disulfoton
18	Vamidothion

III. PESTICIDES RESTRICTED FOR USE IN THE COUNTRY

Sl.No.	*Name of Pesticides*	*Details of Restrictions*
1.	Aluminium Phosphide	The Pest Control Operations with Aluminium Phosphide may be undertaken only by Govt./Govt. undertakings/Govt. Organizations/pest control operators under the strict supervision of Govt. Experts or experts whose expertise is approved by the Plant Protection Advisor to Govt. of India except [1]Aluminium Phosphide 15 per cent 12 g tablet and [2]Aluminum Phosphide 6 per cent tablet. *[RC decision circular F No. 14-11(2)-CIR-II (Vol. II) dated 21-09-1984 and G.S.R. 371(E) dated 20th may 1999]. [1]Decision of 282nd RC held on 02-11-2007 and, [2]Decision of 326th RC held on 15-02-2012.* The production, marketing and use of Aluminium Phosphide tube packs with a capacity of 10 and 20 tablets of 3 g each of Aluminium Phosphide are banned completely. (S.O.677 (E) dated 17thJuly, 2001)
2.	Captafol	The use of Captafol as foliar spray is banned. Captafol shall be used only as seed dresser. (S.O.569 (E) dated 25thJuly, 1989) The manufacture of Captafol 80 per cent powder for dry seed treatment (DS) is banned for use in the country except manufacture for export. (S.O.679 (E) dated 17thJuly, 2001)
3.	Cypermethrin	Cypermethrin 3 per cent Smoke Generator, is to be used only through Pest Control Operators and not allowed to be used by the General Public. [Order of Hon,ble High Court of Delhi in WP(C) 10052 of 2009 dated 14-07-2009 and LPA-429/2009 dated 08-09-2009]
4.	Dazomet	The use of Dazomet is not permitted on Tea. (S.O.3006 (E) dated 31st Dec, 2008)
5.	Diazinon	Diazinon is banned for use in agriculture except for household use. (S.O.45 (E) dated 08th Jan, 2008)
6.	Dichloro Diphenyl Trichloroethane (DDT)	The use of DDT for the domestic Public Health Programme is restricted up to 10,000 Metric Tonnes per annum, except in case of any major outbreak of epidemic. M/s Hindustan Insecticides Ltd., the sole manufacturer of DDT in the country may manufacture DDT for export to other countries for use in vector control for public health purpose. The export of DDT to Parties and State non-Parties shall be strictly in accordance with the paragraph 2(b) article 3 of the Stockholm Convention on Persistent Organic Pollutants (POPs). (S.O.295 (E) dated 8th March, 2006) Use of DDT in Agriculture is withdrawn. In very special circumstances warranting the use of DDT for plant protection work, the state or central Govt. may purchase it directly from M/s Hindustan Insecticides Ltd. to be used under expert Governmental supervision. (S.O.378 (E) dated 26thMay, 1989)
7.	Fenitrothion	The use of Fenitrothion is banned in Agriculture except for locust control in scheduled desert area and public health. (S.O.706 (E) dated 03rdMay, 2007)

Sl.No.	*Name of Pesticides*	*Details of Restrictions*
8.	Fenthion	The use of Fenthion is banned in Agriculture except for locust control, household and public health. (S.O.46 (E) dated 08th Jan, 2008)
9.	Methoxy Ethyl Mercuric Chloride (MEMC)	The use of MEMC is banned completely except for seed treatment of potato and sugarcane. (S.O.681 (E) dated 17thJuly, 2001)
10.	Methyl Bromide	Methyl Bromide may be used only by Govt./Govt. undertakings/Govt. Organizations/Pest control operators under the strict supervision of Govt. Experts or Experts whose expertise is approved by the Plant Protection Advisor to Govt. of India. [G.S.R.371 (E) dated 20thMay, 1999 and earlier RC decision]
11.	Methyl Parathion	Methyl Parathion 50 per cent EC and 2 per cent DP formulations are banned for use on fruits and vegetables. (S.O.680 (E) dated 17thJuly, 2001) The use of Methyl Parathion is permitted only on those crops approved by the Registration Committee where honeybees are not acting as a pollinators. (S.O.658 (E) dated 04th Sep., 1992.)
12.	Monocrotophos	Monocrotophos is banned for use on vegetables. (S.O.1482 (E) dated 10thOct, 2005)
13.	Sodium Cyanide	The use of Sodium Cyanide shall be restricted for Fumigation of Cotton bales under expert supervision approved by the Plant Protection Advisor to Govt. of India. (S.O.569(E) dated 25thJuly, 1989)

Protocols of Seed Treatment of Vegetable Crops through Hot-Water

Crops	*Temperature (°C)*	*Time (Minutes)*
Brussels sprouts	50	25
Broccoli	50	20
Cabbage	50	25
Carrot	50	20
Cauliflower	50	20
Celeriac	48	30
Celery	48	30
Chines cabbage	50	20
Coriander	52.5	30
Cucumber	50	20
Eggplant	50	30
Kohlrabi	50	20
Lettuce	48	30
Mint	47	10
Mustard	50	15

Crops	*Temperature (°C)*	*Time (Minutes)*
Spinach	50	25
Onion	46	60
Radish	52	30
Sweetpotato (roots)	46	65
Tomato	50	25
Turnip	50	20
Yam(tuber)	46	30
Okra	60	30
Tea	50	20
Sugarbeet	55	30
Pepper	10	10
Watermelon	51	10
Pumpkin	50	10
Turmeric	50	30
Guar	50	10
Bean	48	25
Betlevine	51	10
Pea	55	15
Potato	50	20
Cowpea	52	15
Ginger	50	30

List of Biological Control Agent

Sl.No.	*Name of Parasites/Predator*
1	*Trichogramma sp*
2	*Trichogrammatoidae armigera*
3	*Compoletis choridae*
4	*Carcelia illota*
5	*Eriborus argenteopilosus*
6	*Hyposoter didynotor*
7	*Palexorista laxa*
8	*Gonoipthalmus halli*
9	*Telenomus remus*
10	*Chelomus helioptae*
11	*Cotesia colemani*
12	*Microplitis sp*
13	*Charops obtuse*
14	*Euplectus Sp*
15	*Peribaea orbata*

Sl.No.	*Name of Parasites/Predator*
16	*Phaneraotoma sp.*
17	*Chrysonomyia appannai*
18	*Coccinellids*
19	*Syrphids*
20	*Spiders*
21	*Carabids*
22	*Stophylinids*

Chapter 33

Indian Plant Pathologist

The science of plant pathology in modern period in this country is rather not very old. Even when humans lived as hunters or nomads and their food consisted only of meat or leaves, fruit, and seeds. Which they picked wherever they could find them, plant diseases took their toll on hunted animals and on humans. The development of this subject started with the study of mycology. The first study of pathology started in Indian universities that were established in 1857 at Calcutta, Bombay and Madras emphasized collection, identification, characterization and taxonomy of fungi. Up to 1930, there was more attention to the study of fungi, bacteria, virus, MLOs and Nematode than to disease caused by them. Many scientists are contributed in plant pathology are mentioned here such as:

1. D.D. Cunnigham (1850-1875)

Started identification of fungi in this country.

- ✰ Cunnigham was found a species *Gymnosporangium* fungus, and named *G. cunninghamianum.*
- ✰ His studied the spread of bacteria and the spores of fungi through the air and also made studies on cholera.
- ✰ Cunningham made a special study of rusts and smuts.
- ✰ Worked and collected the members of mucorales and uredinales.
- ✰ Identified the causal organism of red rust of tea in Assam caused by *Cephaleurous virescens.*

2. A. Barclay

- ☆ Barclay began his critical work on the rust near Shimla, and during his following 7 years.
- ☆ He studies the life histories of many of the Himalayan Uredinales.

3. K.R. Kirtikar

- ☆ Kirtikar was the first Indian scientist who collected and identified fungi.
- ☆ Kirtikar was collect and identified some of mushrooms.He wrote a Book-Indian Medical Plants

4. E. J. Butler (1874-1943)

- ☆ He was the father of modern plant pathology and father of Indian Mycology.
- ☆ He worked at IARI for 20 year from 1901 to 1920.
- ☆ Butler stayed in India for about 20 year and besides training several Indian scientists, studies wilt of cotton and pigeon pea, disease of rice, toddy palm, sugarcane, potato and cereal rust.
- ☆ He was the founder and first director of Imperial Mycological Institute, England (1920-35).
- ☆ Monograph: *Pythiaceous and allied fungi.*
- ☆ He wrote a Book: Fungi and Disease in plants (1918).

5. B.B. Mundkur

- ☆ He worked as Asst. Mycologist in the cotton research scheme at Dharwar, and after that pointed as Asst. Mycologist in (IARI), Pusa, Bihar.

- ☆ He worked on the control of cotton wilt disease through varietal resistance.
- ☆ He was also responsible for the identification and Indian classification of large number of smut fungi.
- ☆ He started Indian Phytopathological Society in 1948 with its published a ***Journal Indian Phytopathology*** published about 80 papers.
- ☆ He also authored a textbook: **Fungi and plant diseases,** and the monographs 'Genera **of Rusts'.**

6. J.F. Dastur (1886-1971)

- ☆ Born on 10 sep. 1886 at Bhavnagar; M.Sc. in 1909 Bombay university, Entered government.
- ☆ Service in 1910 as a first Asst. to the Imperial Mycologist and appointed as Imperial Mycologist and Head of Plant Pathology, IARI, New Delhi.
- ☆ Dastur a colleague of Butler, was the first Indian plant pathologist. Who is credited with the detailed study of fungi and plant diseases? His special field of study was the genus *Phytophthora* and *Corticium* diseases of castor and potato caused by its species.
- ☆ He established *Phytophthora parasitica* from castor. He has published about 36 original papers and memoirs

7. S.L. Ajrekar

- ☆ Ajrekar was a professor of botany in Institute of science, Mumbai (1923-28) and science institute, Gujarat College, Ahmedabad (1928-38). Ajrekar studied wilt of cotton, *Mucor, Aspergilus*, smut of sugarcane, and ergot of sorghum.
- ☆ Ajrekar was an Editor of Journal of the Indian Botanical Society.
- ☆ The Maharashtra Association for the cultivation of Science named its herbarium 'Ajrekar Mycological Herbarium' in his honour.

8. K.D. Bagchee (1898-1973)

- ☆ Born in January 1898. Obtained M.Sc. degree in Botany from Calcutta University and D.Sc. and D.I.C. from London University.
- ☆ Started career as Mycologist in 1927 at the Forest Research Institute, Dehradun and did pioneering research on mycology and Forest Pathology.
- ☆ Special reference to biology of rust fungi on conifers, soil borne diseases and their control.

- ☆ After retirement in 1956 from the (F.R.I), appointed Director, Botanical Research, J and K State where he initiated and organized work on botanical research and surveyed diseases of forest plants in the state.

9. Dr. R. Prasad (1907-94)

- ☆ Dr. Prasada was born on 11 Dec., 1907 at Chandausi, in (U.P). He was educated at Agra College, taking B.Sc., M.Sc. and D.Sc. Degrees of Agra University.
- ☆ In 1946 Dr. Prasad joined the IARI, New Delhi and initiated work on the epidemiology of cereal rust in association with Dr. K. C. Mehta.
- ☆ He had also been associated since 1935 with the breeding of rust resistant varieties of wheat.
- ☆ Dr. Prasad has been closely connected with the Indian Phytopathological Society. He was editing the journal Indian Phytopathology for six year as Chief-Editor.

10. S.N. Das Gupta (1902-90)

- ☆ Born at Jalpaigury on 4th Nov. 1902 Educated at Bangabasi College, presidency college Calcutta passed M.Sc. (Botany).He awarded Ph.D. and D.Sc. Degrees and D.I.C. of London University.
- ☆ He started his career as Reader in Botany at the University of Lucknow (1934) and became Professor and (H.O.D) Botany, in 1949, and Dean, faculty of science in 1954.
- ☆ He developed a school of Plant Pathology including virology and microbiology at Lucknow University.
- ☆ Conducted researches in sanitation of fungi, nutritional requirement of fungi, mango necrosis, and deficiency disease of mango, aquatic fungi and medical mycology.
- ☆ Dasgupta studies on the controversial mango black tip disease.
- ☆ Dasgupta studied the role of enzymes in pathogencity

11. M.J. Thirumalachar (1914-99)

- ☆ Born in sep. 1914 at Bangalore, received the B.Sc. Botany, and D.Sc. From the University of Mysore; worked under the guidance of Dr. B.B. Mundkur.
- ☆ He created over 20 new genera and 300 new species of fungi.
- ☆ Thirumalachar conducted exhaustive study of smuts, rusts and *Cercospora etc.* He studied their taxonomy, cytology and life-cycles.
- ☆ He successfully introduced the use of antibiotics in plant disease control. Ex. Aureofungin and streptocycline.
- ☆ While in USA, he isolated a new genus, *Geotrichum marinum,* which can grow on the oil, carbon and nitrogen source.

12. Prof. M.K. Patel (1899-1967)

- ☆ Prof. Patel laid the foundation of phytobacteriology in India.
- ☆ He established a school of Plant Bacteriology at college of agriculture, Pune and first described a new sp. *Xanthomonos campestris pv. uppali* in 1948, from the host *Ipomea muricota.*
- ☆ He first documented that black spot of mango is caused by *Pseudomonas mangiferaeindiceae.*
- ☆ He described more than 30 bacterial diseases from India.
- ☆ Such 14as- Bacterial leaf spot on *Vernonia cinerea* in 1968.
- ☆ 15 Bacterial leaf spot of *Cyamopsis tetragonoloba* in 1953.17

13. Abrar M. Khan (1919-2004)

- ☆ He was prof. and Head, Department of Botany at (A.M.U),Aligarh. He obtained a Ph.D. with Prof. Stakman.
- ☆ He worked the development of *Sphareo-theca* and *Erisiphe* infecting cucurbits in tropical climate.
- ☆ In India, he pioneered research on phytonematology. He was also considered as the father of Nematology in India and initiated the Nematological Society of India.

14. Dr. H.K. Saksena (1922-2007)

- ☆ Born on 21 Dec. 1922 at Sultanpur; M.Sc. in Botany, (AU) 1944; Associate IARI in mycology and plant pathology.
- ☆ Started his career in 1952 as lecturer in Botany, (AU).
- ☆ In 1953, he joined as Assistant Prof. of Plant Pathology and continued as Prof. and Head, Department of Plant Pathology since1961 at Govt. Agri. College of Kanpur.
- ☆ Associated with the Indian Phytopathological Society since its inception as a Charter Member, Councilor, Vice President, President and Editor in Chief.
- ☆ His main contribution have been on the parasitism and saprophytic survival of plant pathogenic fungi, and on the taxonomy, cytology, and physiology of parasitism.
- ☆ His main study on *R. solani* and control of *Rhizoctonia* group of fungi.

15. Prof. R.S. Singh (1927)

- ☆ Dr. Singh served as Prof. and Head, Department of Plant Pathology at Pantnagar.
- ☆ He worked on biological control and ecology of soil borne pathogen.
- ☆ His book: "Plant diseases" (1963) is referred as Bible of Plant Pathology in India.
- ☆ Principles of Plant Pathology
- ☆ Diseases of Vegetable Crops
- ☆ Diseases of Fruit Crops
- ☆ Plant Disease Management and 'A Handbook of Economic Nematology'.

16. Prof. R.K. Grover (1931-2007)

- ☆ He worked mainly at Haryana Agriculture University, Hisar, on fungicidal control of plant diseases specially caused by *Rhizoctonia* diseases of crop plants.
- ☆ He worked Chemical control of root rot of cow pea and Studies on *Rhizobium*-fungitoxicant interaction in cow pea.
- ☆ He introduced the concept of environmental factors of soil and aerial parts on the effectiveness of fungicides.

17. Dr. B.B. Nagaich

- ☆ Dr. Banke Bihari Nagaich was born 11th Aug. 1931 at District Hamirpur (U. P.). In the year 1955 are obtained M.Sc. (Ag.) Degree in Plant Pathology from Agra University.
- ☆ Later he obtained his Doctral Degree from the University Of Illionis (US). During his stay at Illionis he was assisted by Prof. Thornberry in

survey and surveillance work on plant viruses.

- Dr. Nagaich joined the IARI at its Regional Station at Shimla in 1958.
- He joined the CPRI, Shimla where he worked as a virologist and as head, Division of Plant Pathology.
- In Oct.1975, Dr. Nagaich took over as the Director of CPRI, Shimla(1975-82) and retired from this post in 1982.
- In 1980-81, he was the first Editor-in –Chief of the journal of Indian Potato Association.
- Dr. Nagaich is also the life member of Indian Phytopathological society. He was the President of IPS in 1982.

18. Dr. Y.L. Nene (1936)

- Yeshwant Laxman Nene was born in Gwalior, Madhya Pradesh in 1936.
- He did his undergraduate education in Agriculture College at Gwalior and post-graduate education in plant pathology at the Agricultural College, Kanpur.
- He started his professional career in 1960 as Assistant Professor and the founder of the Department of Plant Pathology, (GBPUA and T) Pantnagar, U.K., and India.
- In 1989, he was appointed the Deputy Director General of ICRISAT, Hyderabad.
- Dr. Nene, while working at GBPUA and T, solved the mystery of khaira **disease of rice** and established that it was caused by zinc deficiency.
- At ICRISAT, he developed multiple-disease resistance-screening techniques and disease-resistant varieties of pulses.
- Authored the books ***"Fungicides in plant disease control"*** and ***"The pigeonpea"***.
- It was on 30 Dec. 1994 that he established the AAHF and a Trust, (AAHFT). He was elected as the first Chairman of the AAHFT.

19. Dr. B.S. Mehrotra

- ✰ Born in Allahabad in 1927, obtained his M.Sc. Degree, Botany Deptt. (AU).
- ✰ He did his Ph.D. under the guidance of Dr. R.K. Saksena from (A.U).
- ✰ He made excellent contributions on taxonomy of fungi particularly of *Aspergilli* and *Penicillia.*
- ✰ He retired from Head, Department of Botany, (AU).
- ✰ He has written a book **'The Fungi'.**

20. A.N. Mukhopadhyay (1940)

- ✰ He did his Ph.D. from B.H.U., He served as professor. and Head Deptt. of Plant Pathology at Gujarat Agriculture University in 1982.
- ✰ He is known for his work on bio-control of plant disease in Pantnagar University (1984).
- ✰ He is credited with the publication of four edited volumes entitled with ***Plant Disease of International Importance***

21. C. Manoharachary (1944)

- ✰ He was prof. of Botany, Osmania University, Hyderabad, and now Emeritus professor there.
- ✰ The research conducted in bio-diversity of fungi, fungal 'bio-technology, bio-pesticides, plant protection, bio-fertilizer technology, Mushroom cultivation and advances areas received recognization at National and International level.
- ✰ He contributed extensively on the various aspects of soil fungi, Aquatic fungi, forest fungi.
- ✰ Executive editor, Indian Journal of Microbial Ecology since 1991.
- ✰ He has been awarded **"E.K. Janaki Ammal National Award"** for taxonomy by the Ministry of Environment and Forestry, Government of India on 10 Oct. 2006.
- ✰ Research publication-400

  National -320

  International - 80

  Books - 12

- ☆ Author a Monograph "Soil Fungi from Andhra Pradesh"(1990)
- ☆ Author Text Book:- "Botany-Lower Plants"(1993), "Fungi"(1995), "Botany"(1996), "Cultivation of Mushrooms"(1996).

22. Dr. Anandraj

- ☆ Born in Karnataka, India. Completed M.Sc. From University of Mysore, and Ph.D. from University of Calicut.
- ☆ He was Senior Scientist (1984-1998), Principal Scientist, (1998-2005), Head, Division of Crop Protection, (IISR) Calicut, Kerala, (2005-2006).
- ☆ Director, Indian Institute of Spices Research, Calicut (July, 2011 onwards). His research worked on Plant Pathology- Epidemiology, Soil borne diseases especially *Phytophthora,* Biological Control

Disease Terminology

Abiotic: Non living disease causing agent.

Alternate host: One of the other kind of plant variety where pathogen survive and complet its life cycle.

Appressorium: Swollen tip of a hypha or germtube of the fungus that initiate the penetration process in host tissue.

Bacterial slime: A gooey or dried mass of bacterial cells that oozes out of plant tissues.

Bacterium: A single: celled, microscopic organism having a cell wall and reproduces by cell division.

Biological control: Any living organism that inhibit the growth of other microorganisms

Biotic:. Living organisms that survive on other living organism with causing harm

Blight: Rapid, extensive discoloration, wilting, and death of plant tissue.

Blotch: A large superficial and irregular with different shape and size on leaves, shoots, or fruit.

Canker: A dead place on the bark and cortex of twigs, stems, or trunks, often discolored and either raised or sunken.

Chemotherapy: Control of plant disease with chemical that are absorbed and tranlocated in plant tissues

Chlamydospore: A thick walled asexual spore formed by the mycelium of the fungus

Chlorosis: Form of yellowish: white or gray color of plant parts due to destruction of chlorophyll.

Conidiophores: A specialized structure of hyphae those are producing conidia

Conidium: Uncleavage asexual spore of fungus produced on the conidiophores

Culture: Growth of micro: organism on artificial food materials and maintained for long term on same

Cyst: An encysted zoospore of oomyceties or the swollen, *e.g.*: containing female body of certain nematodes; it can be seen on the outside of infected roots.

Damping off: Death of seedling from the collar region or surface of the soil

Dieback: Progressive death of shoots, branches, or roots, generally starting at the tips.

Disease cycle: The chain of events involved in disease development such as survival, penetration, infection and spread.

Disease: Any malfunctioning of host cells and tissues at results from continuous irritation by living (biotic) agents, such as fungi, bacteria, nematodes, or viruses, or by non: living (abiotic) factors, such as nutrient deficiencies or water stress.

Disinfection: Any method either physical or chemical made to free a plant, plant organs and tissues from disease infection

Downy mildew: A plant disease in which the sporangiophores and spores of a fungus produce downy growth on the underside of leaves.

Dwarfing: The underdevelopment of any plant organ. Or reduced size of plants

Enation: Epidermal outgrowths on leaves or stems.

Epidemic: A wide spread of disease among the population of plants in stipulated time

Epidemology: The study of factors affecting the outbreak and spread of infectious diseases

Epinasty: An abnormal downward: curling growth or movement of a leaf, leaf part, or stem.

Eradication: Elimination of the pathogen after its established by chemical or any other substances

Etiology: Study of causal agent of a disease

Etiolation: Development of yellow, long, spindly growth as a result of insufficient light.

Exclusion: A process by which certain plants are not allowed to enter in an area in order to protect against pests and diseases from one area to other areas.

Facultative parasites: Pathogen having the ability to absorbing nutrients from living tissues

Fertilization: The sexual union of two nuclei

Fruiting body: A fungal structure that contains or bears spores.

Fumigant: A taxic gas or volatile substance that are used for destruction of pathogen from soil or certain areas

Fumigation: Methods of application of fumigant

Fungicide: A toxic compound that kill the fungi.

Fungus: A plant organism with no chlorophyll that reproduces via spores and usually has thready growth. Examples are molds, yeasts, and mushrooms.

Gall: An abnormal, localized swelling on leaf, stem, or root tissue of plants due to infection of pathogen

Germtube: Germinating fungus spore by forming special structure

Gumosis: Formation of excess gum by plant tissue

Habitat: The natural place of germination of an microorganism

Haustoria: Simple or modified hyphal branches of fungus that grow into a plant cell and absorb food and water.

Host: A plant affected with a disease. Or living organisms parasitized for absorbing food materials

Hypha: A single filament of a mycelium of fungus

Immunity: A relationship between a plant and pathogen in which the plant does not infected by the pathogen.

Incubation: A period of development during which a pathogen changes to a form that can penetrate or infect a new host plant. Some fungi, for instance, grow a structure called a penetration peg that can grow through a plant's cell walls and produce the symptoms.

Infection: Pathogen has invaded plant tissue and established a parasitic relationship between host and pathogen.

Inoculation: The transfer of a pathogen to a host plant's tissue.

Inoculum: The disease: transmitting parts of a individual pathogen that infect plants.

Integrated control: Use of all available control measures of a disease or of all diseases with the least cost and the least damage of the environment

Intercellular: Penetration hyphae between cells of the host

Intracellular: Penetration of the hyphae within the hast cells

In vitro: Growth of the pathogen culture outside the host

Invivo: Growth of the pathogen in the host

Latent infection: Infected pathogen does not show any symptoms of the infection in the host

Leaf spot: A self limiting area of infected host in the form of lesion

Life cycle: The succesessive stage of growth and development of the microorganisms

Mildew: Growth of mycelium and spores of the pathogen as whitish growth on the surface of the host

Mould: Any profuse growth of pathogens on dead or decaying matter or surface of the host tissue

Mosaic: Non: uniform foliage coloration with a more or less distinct intermingling of normal green and light green or yellowish patches.

Mottle: An irregular pattern of distinct light and dark areas.

Mycelia: Masses of fungal threads (hyphae) that compose the vegetative body of a fungus.

Mycorrhiza: Symbiotic association of a fungus with the roots of a host plant

Necrosis: Death of plant tissue.

Nematode: A microscopic roundworm, usually living in the soil; they feed on plant roots and called pathogens.

Nectrotroph: Fungus feeding on only dead organic tissues.

Obligate parasite: A parasites that can grow only on living organisms.

Parasite: An organism living on another organisms for obtaining their foods from the latter

Parasitic seed plant: A higher plant that lives parasitically on other seed plants.

Pathogen: A disease: causing organism.

Pathogenacity: The capability of the pathogen cause the disease

Pathogenacity factors: The factors involve for disease development such as gene, environment *etc.*

Penetration: The initial invasion of a pathogen into host body

Phyllody: A excessive change of normal stem and flower structure into leafy structure.

Phytopathogenic: A microorganism that can incite disease in plants

Phytoplasma: A microscopic, bacteria: like organism that lacks a cell wall. and can not grow in culture

Powdery mildew: A group of fungal mycelium with spore production that causes white: to: gray powdery mycelia on the host's exterior.

Propagule: The part of an organism such as spore, hyphae and a bacterium that may be transmitted and produce the organism

Pustule: Small blister like growth on epidermis with elevation due to spore production underneath and push outward.

Quarantine: A regulation of shipment to plants or plant parts in an area, usually to prevent invasion by disease, insects, nematodes, or weeds through legal legislation.

Resistance: The ability in a host plant that make it exclude or overcome or slow a pathogen's activities.

Resistant: Qualities of host that hinder the development of a given pathogen

Rot: The softening, discolouration,decomposition and destruction of host tissue as a result of pathogen infection.

Russet: Yellowish: brownis or reddish: brown scar are on the skin of fruit's surface.

Rust: Disease showing rusty symptoms on a plant due to Uredinales fungi

Sanitation: The process of removing and burning of sources of plant pathogens from a growing area

Scab: A crust: like disease lesion.

Secondary infection: Any infection occurred by the inoculums as a result of secondary inoculums after invasion of primary inoculums.

Secondary inoculums: Inoculums produced by pathogen within infected tissue during growing periods.

Signs: physical evidence of a disease: causing agent.

Smut: Dark black powdery masses with odorous spores produced by Ustilaginales fungi

Soft rot: Rot of fleshy tissues of vegetable, fruit and ornamental plants due to maceration of the tissues by pathogens enzymes

Soil solarization: Effort to reduce or eliminate the pathogen from the soil with covering the transparent plastic sheet

Sporangiophore: A specialized hypha bearing one or more sporangium.

Sporangium: Container of asexual spore bearing structure which some time work as single spore.

Spore: The reproductive body of fungi and other lower plants, containing one or more cells. (2) A bacterial cell modified to survive in an adverse environment.

Sterilization: The distruction or elimination of pathogen and other living organisms from soil, container *etc.* by means of heat and chemicals.

Susceptible: Absence of inherent ability to overcome the attack of pathogens

Stylet: The lance: like or hypodermic: needle: like mouthpart of a nematode, used to puncture and feed from plant cells.

Symptoms: external and internal physical disease characteristics expressed by a host plant.

Tolerant: The ability of crop plant that will produce a normal yield even if diseased causing agent s are present without causing disease.

Tumour: Uncontrolled overgrowth of tissue.

Vector: An animal are able to transmit the disease causing agent

Virulence: The degree of pathogen cause to disease

Virus: An infectious agent too small to see with a high: magnification microscope; it multiplies only in living cells.

Water: soaked: Lesions that appear wet and dark and are usually sunken and/or translucent.

Wilt: (1) Lack of freshness and turgor and drooping of leaves from lack of water. (2) A vascular disease disruption of a plant's for normal uptake and distribution of water.

Witches' broom: Abnormal brush like development of many weak shoots or mass proliferation of branches.

Yellowing: Plant disease symptoms produced by yellowing and stunting due to pathogen.

References

Abdel M. H, Yamamoto N, Otawa K, Tada C and Nakai Y (2010). Isolation of bacteriumlike substances producing bacteria from finished cattle manure compost and activity evaluation against some food borne pathogenic and spoilage bacteria. *Journal of General and Applied Microbiology,* 56(2): 151-161.

Adejumo T. O, Ikotun T and Florini D. A (2001). Identification and survival of the causal organism of leaf smut disease of cowpea in Nigeria. *Mycopathologia,* 150(2): 85-90.

Adejumo T. O, Ikotun Tand Florini D. A (2000). False smut of cowpea is caused by *Protomycopsisphaseoli*(Ramak and Subraman.) in *Nigeria.* Abstracts of paper and poster presented at the *Second World*cowpea *Research Conference,* Accra, Ghana, 3-9 September, 2005.

Agrios G. N. (1997). Plant pathology. *4th edition, Academic Press.,* London, (UK).

Akai S (1951). Studies on the pathological anatomy of fungus gall of plants. Mem. *Coll. Agri. Kyoto Univ,* 58: 1-60.

AliM. N (2011). Sustainable Agriculture with Low Cost Technologies(SALoCT). School of Agriculture and Rural Development, Ramakrishna Mission Vivekananda University, Belur Math, West Bengal p. 47.

Alison M. G, and Peter R. F (1999). Insulin releasing and insulin like activity of the

Amine, R., Sarker, B. C., Adhikari, S. K., Sultana, S. and Zubair, T. (2013). Effect of some botanical extracts and cow's urine on *Sclerotium rolfsii,* causal agent of foot and root rot of Betel vine. *The International Journal of Engineering and Sciences* (IJES) Vol. 2 Issue 9 pp. 77-82.

Anandaraj M (2016). Emerging technologies for sustainable seed spices production. National seminar "New Dimensional Approches for enhancement of seed spices productivity and profitabltity under era of climate change" ICAR (NRCSS) AJjmer, 2-3 February2016, lec. -2: 18-20.

Angua, J. F., Gardner, P. A., Kirkegaard, J. A. and Desmarchelier, J. M. (1994). Biofumigation: Isothiocyanates released from *Brassica* roots inhibit growth of the take-all fungus. *Plant Soil*, 162: 107-112.

AnjumTehmina, Fatima Sabin, and Amjad Sana (2012). Physiological changes in wheat during development of loose smut. *Tropical Plant Pathology*, 37 (2): 102-107.

Anonymous (1986). Annual Report of Agricultural Research Station. Ananharajupet, A. P. Agri. Univ.

Anonymous (1991). Package of practice. Turmeric. Source: National Research Center for Spices. *Spice India*. 4: 2-5.

Anonymous (1999). *Organic farming News letter*, 5(2): 11-15.

Anonymous (2005). Food and Agriculture Organization Production Year Book, Published by Economic and Social Department, FAO, Rome.

Anonymous (2008). USDA nutrient data base from Wikipedia. The free Encyclopedia.

Anonymous (2011). All India Co-ordinate Research Project on Spices, Annual Report. 2012.

Anonymous (2012). All India Co-ordinate Research Project on Spices, Annual Report. 2012.

Anonymous (2013). General information on bio-fumigation. http/www. Infonet-bio-vision. org/default/ct/237/7/20/2013 pp. 1-2.

Anonymous (2014). Annual report of National Horticulture Board New Delhi, pp. 16-17.

Anonymous (2014) Annual Report, National Horticulture Board Database 2013-14.

Anonymous: Area and production of vegetable crops, 2012-13. *National Horticulture Board* pp. 29. (2013).

Anoop, K. and Suseela Bhai, R. (2014). Evaluation of antagonistic potential of indigenous *Trichoderma* isolates against *Pythium aphanidermatum* (edson) fitz. causing rhizome rot in turmeric (*Curcuma longa* L.). *Journal of science*, 4(2) 99-105.

Ashlesha and Sugha S. K (2008). *In vitro* antifungal bio efficacy of microbes present in panchgavya against major soil borne pathogens. *Plant Disease Research*, 23(1): 68-74.

Ashworth, S. and Harrisson (1983). Evaluation of mulches for use in the home garden. *Hort. Sci.*, 18: 180-182.

Atwal, A. S. and Dhaliwall, G. S: Agricultural pests of south Asia and their management. *Kalayani Publishers, New Delhi, India*, 505 (2007).

Awasthi P. B, Mishra U. S and Pande B. N (1978). A new record of *Protomyces macrosporus* Unger on *Foeniculum vulgare* L. *Curr. Sci.*, 47: 832-824.

Bacigalova K (2004). Contribution to the knowledge of the *Protomycetaceae* in Slovakia. *Thaiszia J. Bot.,* 14: 1-22.

Bacigalova K (2008) *Protomyces buerenianus* (*Protomycetaceae*) a new species for Slovakia. *Biologia,* 63/1: 40-43.

BacigalovaKamila, PiatekMarcin and WolkowyckiMarek (2005). *Protomyces cirsii-oleracei* (fungi, protomycetales) A new species for Poland. *Polish Botanical Journal,* 50(1): 77–82.

Bal SS, Singh J and Dhanju KC (1995). Genetics of resistance to mosaic and leaf curl viruses in chilli (*Capsicum annuum* L.). *Indian J. Virol.* 11 (1): 77-79.

Bala, K., G. Robideau, A. W. A. M. de Cock, Z. G. Abad, A. M. Lodhi, S. Shahzad, A. Ghaffar, M. D. Coffey and C. A. Lévesque. (2010). *Phytopythium* Abad, de Cock, Bala, Robideau, Lodhi and Lévesque, *gen. nov. Persoonia,* 24: 136-137.

Balakrishnan, P. (1997). Bio-ecology of rhizome rot of pathogen(s) of ginger and management. Ph.D. thesis, University of Calicut, Calicut, 178.

Banerjee MK and Kalloo G (1987). Sources and inheritance of resistance to leaf curl virus in *Lycopersicon. Theor. Appl. Genet.* 73: 707-710.

Bansal, R. D., Khatri, H. L. and Sharma, O. P. (1985). Growth, sporulation and spore germination of *Colletotrichum capsici* as influenced by tobacco mosaic virus in *Capsicum annuum* L. *Phytopath.* 113: 36-42,

Bansal, R. D., Khatri, H. L. and Sharma, O. P. (1978). Interaction between tobacco mosaic virus and *Collerotrichun capsici* in *Capsicum annuum* L. *Phytopathol.* 17: 210-21.

Basak A. Band Lee M. W (2005). Efficacy of cow dung in controlling root rot and *Fusarium*wilt of cucumber. *Indian Journal of Plant Pathology,* 23(1 and 2): 81-84.

Basak, A. B., Lee, M. W. and Lee, T. S. (2002). Inhibitive activity of cow urine and cow dung against *Sclerotinia sclertiorum* of cucumber. *Corean Society of Mycobiology;* 30 (3): 175-179.

Beltagi H. S. E. l, Ahmed O. K andDesoukyW. E. l(2011) Effect of low doses dakuairradiationon oxidative stress and secondary metabolites production of rosemary (*Rosmarinus officinalis* L.) callus culture. *RadiatPhys. Chem,* 80(9): 968-976.

Bezharuah, S. (2000). Staple Strains. India Today (November): 62-63.

Bhai, S. R., Sasikumar, B. and A. Kumar. (2013). Evaluation of ginger germplasm for resistance to soft rot caused by *Pythium myriotylum. Indian Phytopath.* 66(1): 93-95.

Bhardwaj L. N. and Shrestha S. M(1985) Efficacy of fungicide application in the control of stem gall of coriander. *Agriculture Ecosystems and Environment,* 13: 319-323.

Bhat S, Kaushal P, Kaur M and Sharma H. K (2014). Coriander (*Coriandrum sativum* L.): Processing nutritional and functional aspects. *African Journal of Plant Science,* 8(1) 25-23.

Bhaudauria, H. (2002). Cow urine-A magical therapy. Vishwa Ayurveda Parishad. *Int J. Cow Sci.* 1: 32-36.

Borah, R. K., and Bordoloi, D. K: Influence of planting time on the incidence of leaf curl virus disease and whitefly population on tomato. *Indian Jour. Virol.* 14(1): 71-73 (1998).

Braun, U., and Cook, R. T. A. 2012. Taxonomic Manual of the Erysiphales (Powdery Mildews), CBS Biodiversity Series No. 11. CBS, Utrecht. -VII

Brayford, D. (1993). The identification of fungi. International Mycological Institute, Kew, Surrey, England. p. 27.

Broadbent L and Cooper AJ. 1964. The influence oftomato mosaic virus on root growth and theannuals pattern of fruit production. Ann ApplBiol, 54: 23-30.

Brown, P. D. and Morra, M. J. (1997). Control of soil-borne plant pests using glucosinolates containing plants. *Advances in Agronomy* 61, 167-231.

Brownlee H. E, Mceuen A. R, Hedger J and Scott I. M (1990) Antifungal effects of cocoa tannin on the witches' broom pathogen *Crinipellisperniciosa. Physiological and Molecular Plant Pathology,* 36: 39-48.

Brunt AA and Kenten RH. (1971). Pepper veinal mottle virus, a new member of the potato Y group from pepers (*Capsicum annuum* L. and *C. frutescens* L.) in Ghana. *Ann Appl Biol,* 69: 215-43.

Brunt AA, Crabtree K, Dallwitz MJ, Gibbs AJ, Watson Land Zurcher EJ. 1996. Plant Viruses Online: Descriptions and Lists from the VIDE Database. Version: 20th August 1996. URL http: //biology. anu. edu. au/Groups/MES/vide/Davis RI, Thomas JE, McMichael LA, Dietzgen RG, Callaghan B, James AP, Gunua TG and Rahamma S. 2002. Plant virus surveys on the island of New Guinea and adjacent regions of northern Australia. *Austra Pl Path,* 31: 385- 390.

Buren G (1915). Die schweizerischen *Protomycetaceen* mitbesonderer Berucksich-tigungihrer Entwicklungsgeschichte and Biologie. *Beitr. Kryptogamenfl. Schweiz,* 5(1): 1-95.

Buren G (1922) WeitereUntersuchungenuber die Entwicklungsgeschichte und Biologieder *Protomycetaceen. Beitr. Kryptogamenfl. Schweiz,* 5 (3): 1-94

Butlar, E. J. (1907). An account of the genus *Pythium* and some chytridiaceae. *Mem. Dept. Agric. India* (Bot, Sci.), 1(5): 70.

Butler E. J (1918) Fungi and disease in plants. *Thacker spink and co., Calcutta,* p. 547.

Butler, E. J. and G. R. Bisby, 1931. The fungi of India p. 6, 153. *Sci. Monogrol.* ICAR, New Delhi. (6): 153.

Chadha Sanjay, Rameshwar, Ashlesha, Saini J. P and Paul Y. S (2012). *Vedic Krishi*: Sustainable livelihood option for small and marginal farmers. *Indian Journal of Traditional Knowledge,* 11(3): 480-486.

Chadha, K. L. (2001). Handbook of horticulture Directorate of information and publications of agriculture. Indian Council of Agricultural Research. New Delhi, p. 882.

Chandler, W. A. (1958). Control of bacterial spot and ripe rot of Pimineto pepper PI Dis Reptr. 42: 652455.

Chandler, W. A. and Dempsey, A. N. (1963). Disinfectant treatments for freshly harvested pepper seed. *PI Dis. Report*. 47: 325-327.

Chandrasekaran, A. and Rajappan, K. (2002). Effect of plant extracts, antagonists and chemicals (individual and combined) on foliar anthracnose and pod blight of soybean. *J. Mycol. Pl. Patho.* 32 (1), 25-27.

Chandrasekaran, A., Narasimhan, V. and Rajappan, K. (2000). Integrated management of anthracnose and pod blight of soybean. *Ann. Pl. Prot. Sci.* 8 (2), 103-105.

Charest M. H, Beauchamp C. J and Antoun H (2005). Effects of the humic substances of deinking paper sludge on the antagonism between two compost bacteria and *Pythium ultimum. FEMS Microbiology Ecology* 52(3): 219-227.

Charron, C. S. and Sams, C. E. (1999). Inhibition of *Pythium ultimum* and *Rhizoctonia solani* by Shredded Leaves of *Brassica* Species. *Journal of the American Society for Horticultural Science,* 124(5): 462-467.

Chattopadhya, S. B. (1997). Diseases of plants Yielding Drugs, Dyes and Spices, New Delhi: Indian Council of Agric. Res.

Chattopadhyay, S. B. (1967). Disease of plants yielding drugs, Dyes and Spices, New Delhi, *Indian Council of Agric. Res.,* p. 345

Chet, I. and Inbar, J. (1994). Biological control of fungal pathogens. *Applied Biochemistry and Biotechnology,* 48: 37-43.

Chong J, Baltz R, Schmitt C, Beffa R and Saindrenan P (2002). Downregulation of pathogen-responsive tobacco UDP-Glc: phenylpropanoidsglucosyltransferase reduces scopoletin glucoside accumulation, enhances oxidativestress, and weakness virus resistance. *PlantCell,* 14: 1093-1107.

Chowdhury, E. K., Hasan, M. M., Mustarin, K., Hasan, M. S. and Fancy, R. (2009). Efficacy of different fungicides in controlling rhizome rot of ginger. *J. Agrofor. Environ.,* 3(1): 179-181.

Chowdhury, S. (1957). Studies on the development and control of fruit rot of chillies. *Indian Phytopalh.* 10: 55-62.

Chung Y. R, Kim C. H, Hwang I and Chun J(2000). *Paenibacillus koreensis* sp. *nov.,* a new species that produces an iturin like antifungal compound. *International Journal ofSystematic and Evolutionary Microbiology,* 50(3): 1495-1500.

Chung, W. C., J. W. Huang, H. C. Huang and Jen, J. F. (2002). Effect of ground *Brassica* seed meal on control of *Rhizoctonia* damping-off of cabbage. *J. Plant Pathol.,* 24: 211-218.

Cohen, S., Kern, J., Harpaz, I., and Ben J., R: Epidemiology studies of the tomato leaf curl virus (TYLCV) in Jordan Valley, Israel. *Phytoparasitica* 16(3): 259-270 (1992).

Cortes-Eslava J, Gomez-Arroyo S, and Villalobos-Pietrini R (2004). Anti-mutagenicity of coriander (*Coriandrum sativum*) juice on the mutagenesis produced by plant metabolites of aromatic amines. *J. Toxicol. Lett*, 153: 283-292.

Crover, R. K. and Bansal, R. D. (1970). Seed-borne nature of Colletotrichum capsici in chilli seeds and its control by seed dressing fungicides. *Indian Phytopath*. 23: 664-668.

Czaczyk K, Trojanowska K and Stachowiak B (2000). Antifungal activity of *Bacillus* sp. isolated from compost. *Folia Microbiology*, 45(3): 552.

Czech-Kozlowska M and Krzywanski Z (1984). Phenolic compounds and the polyphenol oxidase and peroxidase activity in callus tissue culture-pathogen combination of red rasberry and Didymellaapplanata (Niessl.) Sacc. *Phytopath. Z*, 109: 176-182.

Dabbas M.R, Singh D.P. and Prakash H.G (2009). Management of Stemgall of Coriander through IDM Practice. *International Journal of Plant Protection*, 2(2): 55-257.

Daiho, L., Upadhyay, D. N. (2004). Screening of ginger genotypes from North-Eastern States against rhizome rot of ginger. National Agricultural Technology Project Final Report (2001-2004), Nagaland University, Nagaland. 14.

Dake, G. N. and Edison, S. (1988). Survey for disease incidence in major ginger growing areas of Kerala during 1984 and 1985. *J. Plantn. Crops* 16: 55-57.

Darughe F, Barzegar M, and Sahari M. A (2012). Antioxidant andAnti-fungal activity of Coriander (*Coriandrum sativum* L.) essential oil. *Int. Food Res. J*, 19(3): 1253-1260.

Das, K., Tiwari, R. K. S. and Shrivastava, D. K. (2010). Techniques for evaluation of medicinal plants as antimicrobial agent. Current methods and future trends. *Journal of Medicinal Plants research*; 4 (2): 104-111.

Dator, V. V. and Mayee, C. D. (1981). *Phytopathometry*, Marathwad Agricultural University, Parabhani. p. 95.

De, B. K., Saha, A. K., and. Nath, P. S: Influence Of meterological factor on the buildup of white fly (*Bemasia tabaci*) vector population and incidence of chilli leaf curl virus disease in the plains of west Bengal. *J. Mycopathol. Res* 43 (2): 195-201(2005).

De, R. K., Mukhopadhyay, A. N. (1994). Biological control of tomato damping off with *Gliocladium virens*. *J. Biological control*, 8(1): 34-40.

Deadman, M., Al-Hasani, H, Al-Sa'di, A. (2006). Solarization and biofumigation reduce *Pythium aphanidermatum* induced damping off and enhance vegetative growth of greenhouse cucumber in Oman. *J. Pl. patho.*, 88(3): 335-337.

Dhankar S, Kaur R, Ruhil S, Balhara M, Dhankhar S, and Chhillar A. K (2011). A review on Justiciaadhatoda A potential source of natural medicine. *Afr. J. Plant Sci*, 5(11): 620-627.

Dick, M. W. (1990). *Keys to Pythium.* Department of Botany, School of Plant Sciences, University of Reading, Reading, U. K.

Dickson, B. T. (1925). Colletotrichum v Vrrmicularia. Mycologia. 17: 213-21 7.

Dimri B. K, Khan, M. N. A. and Narayan M. R (1976). Some promising selection of Bulgaria coriander (*Coriandrum sativum* L) for seed and essential oil with a note on cultivation and distillation of oil. *Indian Perfumer,* 20: 13-21.

Dixit D, Srivastava NK, Kumar R, Sharma S (2002). Cultivar Variation in Yield, Metabolism Translocation and Partitioning of 14CO2 Assimilated Photosynthate into Essential Oil and Curcurmin of Turmeric (*Curcuma longa* L.). *J Plant Biol* 29: 65-70. 4.

Dogra S. (2006). Anti-fungal potential of panchgavya against some soil borne pathogens. M. Sc. Thesis, Department of Plant Pathology, CSK Himachal Pradesh KrishiVishvavidyalaya, Palampur, India. p 76.

Dohroo, N. P., Kansal, S., Preeti Mehta and Neha Ahluwalia (2012). Evaluation of eco-friendly disease management practices against soft rot of ginger caused by *Pythium aphanidermatum. Plant Disease Research* (Ludhiana), 27(1): 1-5.

Dohroo, N. P., Korla and B. N. and Rattan, R. S. (1988). Effect of chemical seed treatment on pre-emergence rot of ginger. *Proc. Nat. Sem. Chillies Ginger and Turmeric.* Hyderabad (A. P.): 153-155.

Draper, N. R., and Smith, H: Applied Regression Analysis (3rd edition). *New York: Wiley* (1981).

Dubey, N. K., Shukla, R., Kumar, A., Singh, P. and Prakash, B. (2010). Prospects of botanical pesticides in sustainable agriculture, *Curr. Sci. India.* 98: 479-480.

Dunne, C. P., Dell, B. and Hardy, G. E. S. (2003). The effect of biofumigants on the vegetative growth of five *Phytophthora* species *in vitro. Acta Hort.* (602): 45-51.

Durairaj, V. (1956). Growth of *Colletotrichum capsici* in pure culture. *J Indian Bio. Sci.* 35: 409413.

Eidi M, Eidi A, Saeidi A, Molanaei S, Sadeghipour A, Bahar M, Bahar K (2012). Effect of coriander seed (*Coriandrum sativum* L.) ethanol extract on insulin release from pancreatic beta cells in streptozotocin induced diabetic rats. *J. Phytother. Res.,* 23(3): 404-406.

Ekbote, S. D. (2005). Management of chili fruit rot caused by *Colltotrichum capsici. J. Mycol. Pl. Path.* 35(1), 183.

Epstein E. (1972). Biochemical changes in terminal root galls caused by anectoparasitic nematode, *Longidorusafricanus:* phenols, carbohydratesandcytokinins. *J. Nematol.,* 4: 246-250

Fageria, MS; Choudhary, BR and Dhaka, RS (2003). Vegetable Crops: Production Technology. Kalyani Publishers, New Delhi-110002.

Fageria, M. S., Choudhary, B. R. and Dhaka, R. S. (2006). Vegetable crops production technology, Vol. II, Kalyani Publisher, New Delhi, pp. 223-227.

Fayzalla, E. A., E. El-Barougy and M. M. El-Rayes, (2009). Control of soil-borne pathogenic fungi of soybean by bio-fumigation with mustard seed meal. *J. Applied Sci., 9: 2272-2279.*

Fisher R. A and Yates (1968). Statistical method for research worker Oliver and Boyd Ltd. *Edin burgh and London,* pp. 10.

Fisher, R. A. and Yates (1968). Statistical method for research worker Oliver and Boyd Ltd. Edin burgh and London. pp. 10.

Frankel E (1995). Nutritional benefits of flavonoids. *International conference on food factors: Chemistry and cancer prevention,* Hamamatsu, *Japan Abstracts,* C62.

Friedman P and Rohde R. A (1977). Phenol levels in leaves of tomato cultivarsinfected with *Pratylenchuspenetrans. J. Nematol.,* 9: 257-259.

Fulzele, S. V., Satturwar, P. M. and Dorle, a. K. (2001). Immunostimulant activity of Cow's urine. *Journal of Immunology* and *Pathology.* 3(2): 87-88.

Galindo A, Jeronimo C, Spaans E, and Weil M (2007). An introduction tomodern agriculture. *Tierra Tropical,* 3(1): 91-96

Gamliel, A., Stapleton, J. J. (1993). Characterization of antifungal volatile compounds evolved from solarized soil amended with cabbage residues. *Phytopathology* 83: 899-905.

Garcia, I., Lora, M. J., Cruz, J. D., Benitez, T., Llobell, A. and Pintor-Tora J. A. (1994). Cloning and characterization of a chitinase CHIT42 cDNA from the mycoparasitic fungus *Trichoderma harzianum. Curr Genet,* 27: 83-89.

GaumanE. A and Dodge C. W (1928). Comparative morphology of fungi. *Mc Grow Hill Book Co. Inc.*: pp. 701.

Germination and in *Vitro* Development of *Protomycopsis* Species. *Caryologia,* 30(3): 313-331.

Geyter E. D, Lambert E, Geelen D, Smagghe G (2007). Advances of saponins as natural insecticides. *Pest Technology,* 1(2): 96-105.

Ghosh, R. and Purkayastha, R. P. (2003). Molecular diagnosis and induced systemic protection against rhizome rot disease of ginger caused by *Pythium aphanidermatum, Curr Sci* 85, 1782.

Ghosh, R., Datta, M. and Purkayastha, R. P. (2006). *P. aphanisdermatum* induced disease resistance in ginger. *Indian J. Exp. Biol.* 44: 68-72.

Gjaerum H. (1964). *Protomycetes, Protomycopsis* and *Taphridiumin* Norway. *NyttMagasin for Botanik,* 12: 19–28.

Goel A. K, Kumar S. and Tayal M. S (1983). Biochemical alteration induced in *coriander sativum* L. by *protomyces macrosporus* Unger. *Indian Botinical Reporter,* 2: 62-64.

Gogoi R, Singh D. Vand Srivastava K. D (2001). Phenols asa biochemical basis of resistance in wheat against Karnal bunt. *Plant Pathology,* 50: 470-476.

Gomathi, V. (2001). Studies on the antifungal effects of plant extracts on *Collerotrichum* spp. infecting *Capsicum annum* L. Ph. D. thesis, Department of Biological Sciences, Pondicherry University, Pondicheny, India. 158

Gopal M, Gupta A, Palaniswami C, Dhanapal R and Thomas G. V (2010). Coconut leaf vermiwash: a bio-liquid from coconut leaf vermin compost for improving the crop production capacities of soil. *Current Science,* 98(9): 1202-121.

Gorawar, M. M., Hedge, Y. R. and Kulkarni, S. (2006). Biology and management of spot of turmeric caused by *Colletotrichum capsici, Journal of Plant Disease Science;* 1(2): 156-158.

Gorawar, M. M., Hedge, Y. R. and Srikant Kulkarni (2005). Chemical control of leaf spot of turmeric caused by *Colletotrichum capsici. Annals of Agri. Bio. Res.* 10: 2, 159-162.

Green S. K. and J. S. Kim (1991). Characteristics and control of viruses infecting peppers: a literature review. Asian Vegetable Research and Development Center. Technical Bulletin No. 18. 60 p.

Grewal, A. S, and Grover, R. K. (1974). Changes in the amino acids, Carbohydrates and capsaicin content of red fruits of chilies infected with *Colletotrichum piperanum. Indian Phytopath.* 27: 603404.

Gupta J. S (1954). Disease appraisal of stem gall of *Coriandrum sativum* L. *Indian phytopath,* 7(1): 53-60.

Gupta J. S (1958). Laboratory assay of certain fungicides, antibiotics and sulpha drugs on the germination of *Protomyces macrosporus* Unger. *Agra Unive. Jour. Res.* (Sci.), 7: 105-112.

Gupta J. S (1962). Pathological anatomy of the floral parts and fruits of coriander affected with *Protomyces macrosporus* Unger *Agra Univ. Jour. Res.,* 11: 307-311.

Gupta J. S and Sinha S (1963a). Biometrical grouping in *Protomyces macrosporus* Unger causing stem gall of Coriander. *Proc. Ind. Acad. Sci.,* 33(4): 507-510.

Gupta J. S and Sinha S (1963b). Some therapeutic control trials on the stem gall disease of coriander. Bull. *Indian Phytopath.* Soc., 1: 75-78.

Gupta J. S and Sinha S (1964). Variation in pathogenicity of *Protomyces macrosporus* Unger. *Proc. Natn. Acad. Sci.,* 34(3): 241-244.

Gupta J. S and Sinha S (1973). Varietal field trial in control of stem gall disease of coriander. *Indian Phytopath.,* 26(2): 337-340.

Gupta P. P, Gupta S. K, Kaushik C. D and Yadava T. P (1985). Biochemical changes in leaf surface washings of groundnut due to tikka disease *Cercosporidium personatum. Indian Phytopath.,* 38: 339-340.

Gupta R. N (1973). Longevity of chlamydospores of coriander gall fungus. *Indian Phytopath.,* 26(3): 581-582.

Gupta R. N (1975). Mineral matter content in coriander leaves and fruits as influenced by stem gall disease. *Indian Phytopath.,* 28: 136-137.

Gupta R. N (1976). Impact of nitrogen, phosphorus and manganese on the stem gall of coriander. *Acta Botanica Indica*, 40(1): 30-35.

Gupta S. K, Gupta P. P, Yadava T. P and Kaushik C. D (1990). Metabolic Changes in mustard due to *Altemaria* leaf blight. *Indian Phytopath.*, 43: 64-69.

Gupta, S. K. and Tripathi, S. C. (2011). Fungitoxic activity of *Solanum torvum* against *Fusarium sacchari. Plant Protection Science*, 47(3): 83-91.

Gupta, V. K., Sharma, R., Singh, S., Jindal, J and Dilawari, V. K. Efficiency of *Bemisia tabaci* (Gennadius) populations from different plant host for acquisition and transmission of cotton leaf curl virus. *Indian journal of Biotech.* 9: 271-275(2010).

Harborne J. B (1973). Phytochemical Methods. *1st ed. Chapman and Hall, London*, 273.

Harsh Deep Yadav, Poonam Yadav, Lokendra Yadav And Prabhat Kumar Yadav (2013). Efficacy of chilli mottle virus disease on root constraints of *Capsicum annuum* L. at Agra, India. *World J. of Applied Sci. and Res.* 57-59.

Hasnat, M., Rajib Bose, Hossain, M. A. Anam, M. M., Kabir, H. (2014). Effect of chemicals, bio-agent plants extract and soil amendments in controlling rhizome rot of ginger. *International J. Natural and Social Sci.*, 1: 01-11.

Haware M. P and Pavgi M. S (1976). Chlamydospore germination in *Protomycopsis* species. Faculty of Agriculture, Banaras Hindu University, India. *Mycopathologia*, 59(2): 105-111.

Haware M. P and Pavgi M. S (1976). Host Range of Two *Protomycopsis*Species from India. Faculty of Agriculture, Banaras Hindu University, India, 272-274.

Haware M. P and Pavgi M. S (1977). Cytology of Chlamydospore

Haware M. P, and Pavgi M. S (1971). Perpetuation of *Protomycopsis* Species Causing Purple Leaf' Spot of Legumes in India. *Ann. Phytopath. Soc. Japan*, 37: 242-248.

Henig-Severa N, Eshelb A, and Neemana G (2000). Regulation of thegermination of Aleppo pine (*Pinus halepensis*) by nitrate, ammonium and gibberellin, and its role in post-fire forest regeneration. *Physiologiaplantarum.* 108: 390-397.

Hidayat HS, Opriana E, Manzila I and Sujiprihati S. (2012). Occurrence of *Chili Veinal Mottle Virus* (ChiVMV) in Indonesia and response of chili *World Journal of Applied Sciences and Research*, 2013, 3(1): 57-59

Horowitz, A. R., Podoler, H., Gerling, D. (1984). Life Table Analysis of the Tobaco Whitefly *Bemisia tabaci* (Gennadius) in Cotton Fields in Israel. *Acta Oecologica Oecologia Applicata* 5(3), 221-233 *Agric. Eco. Environ.*

Horowitz, A. R. (1986). Population dynamics of *Bemisia tabaci* (Gennadius): with special emphasis on cotton field., 17: 37-47.

Howell CR. (1991). Biological control of *Pythium* damping-off of cotton with seed coating preparations of *Gliocladium virens*. *Phytopathology*, 81, 738-741.

Hudge, B. V., Ghugul, S. A. (2009). Evaluation of phytoextracts against *Colletotrichum capsici* causing leaf spot of turmeric. *Annals of Plant Physiology*; 23(2): 276-277.

Isaac S. (1992). Fungal life-style. In: *Fungal plant-Interactions,* Chapman and Hall, London, England, p. 1.

Isabelle Louis, Chew, A. and Lim, G. (1988). Influence of spore density and extracellular conidial matrix on spore germination in *Colletotrichum capsici. Trans. Br. Mycol. Soc.* 91: 694497.

J. Talukdar, A. K. Saikiaand P. Borah (2015). Survey and detection of the diseases of Bhut Jolokia (*Capsicum chinense* Jacq.) in Assam. *Journal Crop and Weed.* 11: 186-192.

Jagtap, G. P. (2012). Control of *Colletotrichum trancatum* causing anthracnose/pod blight of soybean by aqueous leaf extracts, bio-control agents and fungicides. *Scientific Journal of Agriculture.* 1(2) 39-52.

Jagtap, G. P., Mali, A. K. and Utpal D. (2013). Bio efficacy of fungicides, Bio-control agents and botanicals against leaf spot of turmeric incited by *Colletotrichum capsici. African Journal of Microbiology Research;* Vol. 7(18): 1865-1873.

Jain, N. K., Gupta V. B., Garg R. and Silawat (2010). N. Efficacy of cow urine therapy on various cancer patients in Mandsaur District, India - A survey. *Int J Green Pharm;* 4: 29-35.

JaindaikSavita, Thakur preeti and Kumar Vikas (2015). Efficacy of cow urine as plant growth enhancer and antifungal agent. *Advances in Agriculture,* 7: ID 620368.

Janardhanan M and ThoppilJ. E (2004). Herb and spice essential oil. *Discovery publishing House,* New Delhi, 110002: 40-42.

Jayalakshmi, C., Durairaj, P., Seetharaman, K. and Sivaprakasam, K. (1998). Bio-control of fruit rot and die-back of chili using antagonistic microorganisms. *Indian Phytopath.* 51 (2), 180- 183.

Jayanta, T. C. R. (2003). Differential reactions to Colletotrichum leaf spot and genetic variability in some cultivars of turmeric. *Annals of Plant Protection Sciences.* 11: 300-303.

Jeumkya, H. and Byungkook, H. (1998). Influence of inoculum density, wetness duration, plant age and inoculation method and cultivar resistance on infection of pepper plants by *Colletotrichum capsici. Plant Dis.* 82: 1079-1083. 012

Johnston, A. and Booth, C. (eds) (1983). Plant Pathologist's Pocket-Book, 2nd edn. Kew, Surrey, U. K.: Commonwealth Mycological Institute.

Jones, D.R. (2003). Plant viruses transmitted by whiteflies. *European J. Plant Pathol.,* 109: 195–219.

Joseph B and Sankarganesh P (2011). Antifungal efficacy of panchagavya. *International Journal of Pharm Tech Research,* 3: 585-588.

Joseph Philip and Nair, P. C. S. (1981). Field reaction of turmeric types to important pests and diseases. *Indian Cocoa Arecanut and Spices Journal.* 4: 107-109.

Joseph, L. and Sivaprasad. (2000). Biological control of rhizome rots of ginger. *In*: Proceedings of Indian Phytopathological Soceity Golden Jubilee and International Conference on Integrated Plant Disease Management for Sustainable Agriculture, Vol. I. at New Delhi, from November, 11-15, 1997, pp. 339-340.

Joshi, C. G. and Wangikar, P. D. (1978). Efficacy of fungicides against *Colletotrichum capsici*. *Indian Phytopath.* 31: 222-223.

Joshi, D. and Tripathi, H. S. (2002). Cultural, biological and chemical control of anthracnose of urdbean. *J. Mycol. Pl. Path.* 32 (1), 52-55.

Joshi, L. K. and Sharma, N. D. (1980). Diseases of Ginger and Turmeric, Proceedings of the National Seminar on Ginger and Turmeric held at Calicut. April 8-9. pp. 104-109.

Joshi, L. K. and Sharma, N. D. (1982). Diseases of ginger and turmeric. In: Nair M K, Premkumar T, Ravindran P N and Sarma Y R (Eds.). Proceedings, National Seminar on Ginger and Turmeric, Calicut (pp. 104. 117). Central Plantation Crops Research Institute, Kasaragod.

Kai H, Ueda T and Sakaguchi M. (1990) Antimicrobial activity of bark-compost extracts. *Soil Biology and Biochemistry*, 22(7): 983-986.

Kallupurackal, J. A. and Ravindran, P. N. (2002). Turmeric: hints for cultivation. *Spice India* 15: 6-1 1.

Kambar, Y., Vivek, M. N., Manas, M., Kekuda, P. T. R., Nawaz, N. A. S. (2013). Inhibitory effect of cow urine against *Colletotrichum capsici* isolated from anthracnose of chili (*Capsicum annum* L.). *Science Technology and Arts Research Journal.* 2(4): Article in press.

Kar, A. K. and Mahapatra, H. S. (1981). New host records of *Colierotrichum* species from India. *Indian Phytopath.* 34: 219-221

Katan, J., Greenberger, A., Aton, H. and Grinstein, A. (1976). Solar heating by polythene mulching for the control of diseases caused by soil borne pathogens. *Phytopathology*, 66: 683-688.

Kaur, M., Sharma, O. P. and Sharma, P. N. (2006). *In vitro* effect of Trichoderma species on C. capsici causing fruit rot of chili. *Indian Phytopath.* 59 (2), 243-245.

Kaur, S., Singh, J. and Sooch, B. S. (1985). Natural occurrence of different spp, of Collero~rirh~rm attacking *Capsicum annum* L. *Indian Phytopulh.* 38: 755.

Kavyashree, R. (2009). An efficient *in-vitro* protocol for clonal multiplication of Ginger var. varada. *Indian J. of Biotech.*, 8: 228-231.

Keshwal R. L and Khatri R. K (1998). Reaction of some high yielding varieties of coriander to powdery mildew. *J. Myco. and Pl. Patho*, 28(1): 58-59.

Khaing N. S, Khai A. A, Aye K. S and Khaing O. Z (2008) Maintenance of beneficial microbes in vermiwash carrier system and its application in crop. *GMSARN International*, 14(3): 12-16.

Khatso, Kevimeo and N. Tiameren Ao. (2013). Biocontrol of Rhizome Rot Disease of Ginger (*Zingiber officinale* Rosc.). *International Journal of Bio-resource and Stress Management*, 4(2): 317-321.

Kim, J. W. and Park, E. W. (1999). Occurrence and Pathogenicity of *Pythium species* Isolated from Leaf Blight Symptoms of Turfgrasses at Golf Courses in Korea. *Plant Pathol. J.*, 15(2): 112-118.

Kirkegaard, J. A., Wong, P. T. W., Desmarchelier, J. M. and Sarwar, M. (2006). Suppression of soil-borne cereal pathogens and inhibition of wheat germination by mustard seed. Proceedings of the 13th Australian Agronomy Conference, September 10-15, 2006, Australian Society of Agronomy.

Kirkgaard, J. A., Mathiessen, N. P., Wong, T. T. W., Mead, A., Sarwar, M., Smith, B. J. (1999). Exploring the bio-fumigation potential of Brassicas in farming systems. Proceedings of 10th International conference on rapeseed, Canberra, Australia.

Kirtikar K. Rand Basu B. D (1956). Indian Medicinal Plants. *2nd Edition*. Lalit Mohan Basu, Allahabad, Uttar Pradesh, India, III: 2045–2048.

Koche Mina; kothikar, R. B. and Patil, C. U. (2009). Fungicides and botanicals for managing foliar diseases of turmeric. *J. Pl. Dis. Sci.*, 4(1): 68-71.

Konate, K., Kiendrebeogo, M., Ouattara, M. B., Souza, A., Lamienmeda, A., Nongasida, Y., Barro, N., Millobo-rasolodimby, J. and Nacoulma, O. G. (2001). Antibacterial potential of aqueous acetone extract from five medicinal plant used traditionally to treat infectious diseases in Burkina faso. *Current Research of Biological Sciences*. 3 (5): 435-442.

Kulkarni (2011). Etiology, epidemiology and integrated management of rhizome rot of ginger in Karnataka. *Indian Phytopath.*, 64(2): 113-119.

Kulkarni, S., Sagar, S. D. and Lokesha. N. M. (2005b). Integration of soil solarization in the management of rhizome rot of ginger. Paper presented in 'National symposium on modern concepts in plant disease management and their global thrust' and I. P. S. Annual meet (South Zone), Annamalai University, Annamalainagar, T. N. from 29-30 Dec. p. 24.

Kulkarni, S., Sagar, S. D., Hegde, Y. R. and Lokesha. N. M. (2004). *In vitro* screening of botanicals against *Pythium aphanidermatum* and *Fusarium solani* causing rhizome rot of ginger. Paper presented in symposium on Biology, Biotechnology. Epidemiology and management of plant diseases and IPS annual meeting (South zone) held at Gulbarga University, Gulbarga, Karnataka.

Kumar S, Kumar S, Singh M, Singh AK, and Rai M (2006). Identification of sources and inheritance of Pep LCV in *Capsicum*. *Sci. Hort.* 110: 359-361.

Kumar Sanjay (2008). Screening of germplasm and genetics of resistance against leaf curl virur of *Capsicum*. Ph. D. Thesis, Udai Pratap Autonomous College, Varanasi, India.

Kumar, A., Mohanty, A. K., Borah, T. R., and Singh, A. K. (2012). Bio-mulching in organic ginger production under rain-fed conditions in NE hills region. *Indian J. Traditional Knowledge*; 11(3): 487-492.

Kumar, A., Reeja, S. T., Bhai, R. S. and Shiva, K. N. (2008). Distribution of *Pythium myriotylum* Drechsler causing soft rot of ginger. *J. Spices and Aromatic Crops.* 17(1): 5-10.

Kumar, R., Rai, N. and Lakpale, N. (1999). *Orissa Jour. of Hort.* 27: 100-102.

KumarGagan, Yadav S. K, Patel J. S, SarkarA, and Awasthi L. P (2014). Management of Stem Gall Disease in Coriander Using *Pseudomonas* and *Trichoderma* (Bio agents) and Fungicides. *Journal of pure and applied microbiology,* 8: (6).

Kushalappa A. C and Ludwig A (1982). Calculation of apparent infection rate in plant diseases: Development of a method to correct for host growth. *Phytopathology,* 72: 1373-1377.

Kushalappa, A. C. and Ludwig, A. (1982). Calculation of apparent infection rate in plant diseases: Development of a method to correct for host growth. *Phytopathology,* 72: 1373-1377.

Kushwah, S. S., Dwivedi, Y. C. and Jain, P. K. (2013). Effect of mulching materials on growth and yield attributes and enhancing farm income through ginger cultivation under rainfed rice based production system. *Vegetable Science,* 40(1): 96-98.

Kuvalekara A, RedkaraA, Gandheb K and HarsulkaraA (2011). Peroxidase and polyphenol oxidase activities in compatible host-pathogen interaction in *Jasminumofficinale* and *Uromyceshobsoni*: Insights into susceptibility of host. *New Zealand Journal of Botany,* 49(3): 351-359.

Lakra B. S (1990). Effect of depth and amount of inoculum and time of inoculation of chlamydospores of *Protomyces macrosporus* in infectivity and severity of coriander. *Plant Disease Research,* 5(1): 118.

Lakra B. S (1999). Assessment of losses due to stemgall in coriander. *Plant Disease Research,* 14(1): 85-87.

Lakra B. S (2001). Estimation of chlamydospores of *Protomyces macrosporus* in different infected plant parts of coriander. *Plant Disease Research,* 16(1): 94-95.

Lakra B. S (1993). Effect of depth and amount of inoculum and time of inoculation of chlamydospores of *Protomyces macrosporus* in infectivity and severity of coriander. *Plant Disease Research,* 5(1): 118.

Lakra B. S (2000). Management of stem gall of coriander (*Coriandrum sativum*) incited by *Protomyces macrosporus. Indian Journal of Agricultural Sciences,* 70(5): 338-340.

Larkin, R. P. and Griffins, T. S. (2007). Control of soilborne potato diseases using *Brassica* green manures. *Crop Prot.,* 26: 1067-1077.

Latif, M. A., and Akhter, N. (2013). Population Dynamics of Whitefly on Cultivated Crops and its Management. *International Journal of Bio-resource and Stress Management* 4(4): 576-581.

Lewis Y. S (1984). Spices and Herbs in food, Industry. *Food Trade Press Oxpuigton, UK,* pp. 200.

Li, Y.; Mao, L. G.; Yan, D. D.; Liu, X. M.; Ma, T. T.; Shen, J.; Liu, P. F.; Li, Z.; Wang, Q. X.; Ouyang, C. B.; Guo, M. X. (2014). First report in China of soft rot of ginger caused by *Pythium aphanidermatum*. *Plant Disease*, 98(7): 10-11.

Lily V. G and RamadasanA (1979). Changes in phenolic content of coconut leafin relation to the development of leaf rot. *Indian Phytopath*, 32: 112-113.

Lin, L. T., S. S. Chang and L. S. Leu (1971). Soft rot of ginger. *Pl. Prot. Bull. Taiwan*, 12: 54-67.

Lodhi, A. M., Ali Khanzada, M., Shahzad, S. and Ghaffar, A. (2013). Prevalence of *Pythium aphanidermatum* in agro-ecosystem of Sindh province of Pakistan. *Pak. J. Bot.*, 45(2): 635-642.

Lokesh, M. S., Patil, S. V., Gurumurthy, S. B., Palakshappa, M. G., Anandaraj, M. (2012). Solarization and antagonistic organisms for management of rhizome rot of ginger in Karnataka. *International J. Plant Protec.*, 5(2): 195-200.

Lorena T, Calamassi R, Mori B, Mugnai L, and Surico G (2001). *Phaeomoniella chlamydospora* grapevine interaction: histochemical reactions to fungal infection. *Phytopathologia Mediterrranea*, 40: 400-406.

M. Ashfaq, S. Iqbal, T. Mukhtar and H. Shah (2014). Screening for resistance to cucumber mosaic cucumovirus in chilli pepper. *The Journal of Animal and Plant Sciences*, 24(3): 791-795.

M. H. Rashid, K. M. Khalequzzaman, M. S. Alam, S. A. Uddin And S. K. Green (2007). Screening of different sweet pepper lines against cucumber mosaic virus and chili veinal mottle virus. *Int. J. Sustain. Crop Prod.* 2(3): 1-4.

Maheshwari D. K andChaturvedi S. N (1981) Histochemical localization of DNA and histones in the hypertrophied parts of Coriandrum sativum due to stem gall disease. *Indian Phytopath*, 34(4): 488-491.

Maheswari, N. U. Sirchabai, T. P. (2011). Effect of *Trichoderma* species on *Pythium aphanidermatum* causing rhizome rot of turmeric. *Biosciences, Biotechnology Research Asia*, 8(2): 723-728.

Makkar H. P. S, Norvsambuu T. L, Khavatsere S and Becker K (2009). Plant secondary metabolites in some medicinal plants of Mongolia used for enhancing animal health and production. *Tropicultura*, 27(3): 159-167.

Manandhar, J. B., Hadman, C. L. and Wang, T. C. (1995a). Conidial germination and appressorial formation of *Colletorrichum capsici* and *Colletotrichum gloeosporioides* isolates from pepper. *Plant Dis.* 79: 361-366.

Manandhar, J. B., Hadman, C. L. and Wang, T. C. (1995b). Semiselective medium for Colletotrichum gloeosporioides and occurrence of three *Colletotrichum* spp. on pepper plants. *Plant Dis.* 79: 376-379.

Mansangwa, J. I. G., Aveling, P. A. S. and Kritzinger (2013). Screening of plant extract for antifungal against *Colletotrichum spp.* of common bean (*Phaseolus vulgaris* L.) and cowpea (*Vigna anguilata* (L.) Walp. *Journal* of *Agricultural Science.* 151: 482-491.

Masoko, P., Picarid, J. and Eloff, J. N. (2005). Antifungal activities of six South African *Terminalia* species (Combretaceae). *Journal of Ethonopharmacology* 99: 301-308.

Mathai, P. J., Dubey, G. S., Peter, K. V., Saklani, U. D., Singh, N. P. (1977). Pant C-1 and Pant C-2—two new promising selections of chilli *Capsicum annuum* L. *South Indian Hort.* 25, 123–125.

MathivananR, Edwin, S. C, Viswanathan K, and Chandrasekaran D (2006). Chemical Microbial composition and antibacterial activity of modified panchagavya. *International Journal of Cow Science*, 2(2): (retrieved 12. 02. 10 from *Indian Journals. com*).

Mathur S. B (1962). Soil pH in relation to stem gall disease of coriander caused by *Protomyces macrosporus* Unger. *Indian Phytopatho.*, 15: 75-76.

Mathur S. B and Narula P. N (1963). Effect of certain fungicides, antibiotics and synthetic phytohormonesthe germination of chlamydospores of *Protomyces macrosporus* Unger causing stem gall of coriander. *Proc. Ind. Acad. Sci. Sec.*, 33(4): 615-617.

Mathur, Kusum; Ram, D., J. Poonia and Lodha., B. C. (2002). Integration of soil solarization and pesticides for management of rhizome rot of ginger. *Indian Phytopath.* 55(3): 345-347.

Matthiessen, J. N. and Kirkegaard, J. A. (1998). Biofumigation for managing soil-borne pests- progress, pitfalls and prospects. Proceedings of the 6th Australasian Applied Entomology Research Conference (Eds Zalucki M, Drew R and White G). Volume 1: 364-372, University of Queensland.

Mayton, H. S., Olivier, C., Vaughn, S. F. and Loria, R. (1996). Correlation of fungicidal activity of *Brassica* species with allyl-isothiocyanate production in macerated leaf tissue. *Phytopathology* 86: 267-271.

Mc Donald, B. A. and Linde, C. (2002). Pathogen population genetics, evolutionary potential, and durable resistance. Annual Review of *Sclerotium rolfsi*, *Phytopathology*, 71: 1099-1003.

Mc Rae (1911). Efficacy of fungicides for seed treatment against rhizome rot of ginger. *Indian Journal of Plant Pathology*, 19(1 and 2): 39-43.

McRae. W. (1908). *Rep. Agric Sta. E. Bengal and Assam, p.* 30 (1911). Soft rot of ginger in the Rangpur district. *East Bengal Agric. J.* India, 6: 139-146.

Meena, R. Panti, V. and Arora, D. K. (2006). An epidemic of chilli leaf curls disease in Rajasthan. *J. Phytol. Res.* 19(2): 335-336.

Memane SA, Joi MB, and Kale PN (1987). Screening of chilli cultivars against leaf curl complex. *Current Research Reporter*. Mahatma Phule Agricultural University 3(1): 98-99.

Mishra R. S and Pandey V. P (2013). Effect of bio agents and fungicides on the management of stem gall of coriander. Presented in National symposium on "Emerging pollutant and pathogen due to climate change challenges and risk reduction", pp. 82.

Mishra R. S and Pandey V. P (2015). Evaluation of varieties of coriander (*Coriandrum sativam* L.) for resistance to stem gall disease and seed yield. *Curent advances in Agricultural Sciences*, 7(2): 151-153.

Mishra R. S and Pandey V.P. (2015). Screening of *Coriandrum sativum* accessions for stem gall disease tolerance and seed yield. *J. Plant Disease Science*, 10 (2): 256-259.

Mishra R. S., Ramveer and Pandey V. P (2015). Antifungal effect of cow urine extracts against *Colletptrichum capsici* causing leaf spot disease of turmeric, *African Journal of Biotechnology*, 14(20).

Mishra R.S. and Kumar Sandeep (2016). Viability of coriander seeds from *Protomyces macrosporus* infected plants to spread of stem gall of coriander. *CAAS*: 26(02): in press.

Mishra, A. C. and Pandey, V. K. (2009). Effect of mulching on plant growth and rhizome yield in ginger (*Zinziber officinale* Rosc.) in rainfed plateaus of Jharkhand. *International J. Pl. Sci;* 4(2): 385-387.

Mishra, R. S. and Pandey, V. P. (2014). Management of leaf spot of turmeric caused by *Colletotrichum capsici* through fungicides. "National symposium on plant pathology in genomic era and 66th Annual meeting of IPS". 26th to 28th may 2014 Raipur p. 22.

Mishra, R. S. and Pandey, V. P. (2014). Management of soft rot of ginger through Bio-Fumigation, Fungicides and bio-agents. *J. Pl. Dis. Sci.* 9(1): 128-130.

Mishra, R. S. and Pandey, V. P. (2015). Management of *Colletotrichum* leaf spot of turmeric through fungicides. *J. Spices and Aromatics Crops;* 24(2): 153-156.

Mishra, R. S. and Sandeep Kumar (2015). Evaluation of plant extract, bio-agents and fungicides against *Fusarium udum* causing pigeon wilt. *J. Pl. Dis. Sci.* 10(1) 92-98.

Mishra, R. S., Pandey, V. P. and Ankit Tiwari (2015). Screening of genotypes and effect of fungicides against leaf blotch of turmeric. *J. Pl. Dis. Sci.* 10(1) 99-102.

Misra, A. P. and Dutta, K. K. (1963). Studies in anthracnose fungi, 111. A comparative study of two isolates of *Colletotrichum capsici* (Syd.) Butler and Risby. *J Indian Bio Sci.* 42: 74-84.

Misra, A. P. and Mahmood, M. (1960a). Factors affecting the growth of *Collerotrichum capsici. Indian Phytopath.* 13: 12-1 7.

Misra, A. P. and Mahmood, M. (1960b). Effects of carbon and nitrogen nutrition on growth and sporulation of *Colletorrichum capsici* (Syd.) Butler and Bisby. *J Indian Bio Sci* 39: 314-321.

Mix A. J (1925). *Ibid.* 15: 214-222.

Mohamed A. A, Khalil A. A, Beltagi H. E. SE.L. (2010). Antioxidant and antimicrobial properties of kaff Maryam (*Anastaticahiero chuntica*) and doum palm (*Hyphae nethebaica*) Grasas Y Aceites, 61(1): 67–75.

Mohan B (2008). Evaluation of organic growth promoters on yield of fruits crops in *India. J. Organ. Syst.*, 3(1): 23-36.

Mohanty K. C, Dash L, and Swain S. C (2001). Effect *of Meloidogyne incognita* on nodulation and root biochemistry in French bean. *Ann. Plant Protec. Sci.,* 9: 81-86.

Mohanty, I., Senapati, M. R., Jena, D. and Palai, S. (2014). Diversified uses of cow urine. *International Journal of Pharmacy and Pharmaceuticals Science. Vol.* 6 Issue 3: 45-48.

Mordue, J. E. M. (1971). *Glomerella cingulata.* C. M. I. Description of Pathogenic fungi and Bacteria. No. 315.

Muhammad S and Amusa N. A. (2003). *In vitro* inhibition of growth of some seedling blight inducing pathogens by compost inhibiting microbes. *African Journal of Biotechnology,* 2(6): 161-164.

Mukhopadhyay A. N (1968) Studies on stem gall of coriander caused by *Protomyces macrosporus*Unger. Ph. D. Thesis, B. H. U. Varanasi, p. 152.

Mukhopadhyay A. N and Pavgi M. S (1964). Chlamydospore germination of artificial culture of *Protomyces macrosporus* Unger. *Experimentia,* 20(11): 619.

Müller K O, and Börger H (1940). Experimentell euntersuchungenuber die *Phytophthoraresistenzder* kartoffel. *Arb. Biol. Reichsasnstalt Landw Forstw Berlin,* 23: 189–231.

Mundkur B. B (1949). Fungi and Plant Disease. *McMilan and co. Ltd., London.* p. 107-109.

Mundkur B. B (1959). Fungi and Plant Disease. *Macmillan and Co. Ltd., London,* pp. 246.

Muruganandam, V., Suryanarayanan, R. S. and Venkataehalam, S. (1987). Formation of infection structure of *Colletotrichum capsici* as influenced by nitrogen sources. *Curr Sci.* 13: 674475.

Nabavi S. M, Ebrahimzadeh M. A, Nabavi S. F, Hamidinia A, Bekhradnia A. R (2008). Determination of antioxidant activity, phenol and flavonoids content of *Parrotiapersic a*Mey. *Pharmacologyonline,* 2: 560-567.

Nageshwar Rao, T. G., Reddy, B. N. and Sasikumar, B. (1993). Turmeric: production constraints of turmeric in Andhra Pradesh. *Spice India.* 6: 8-11.

Naidu, P. H. (1988). Spices Newsl. Spices Board, Cochin, June pp. 10-13

NaikN, and Sreenivasa M. N (2009). Influence of bacteria isolated from Panchagavya on seed germination and seed vigour in wheat. *Karnataka J. Agric. Sci.,* 22(1): 231-232.

Nakasone A. K, Bettiole W, and Souze R. M. D (1999). The effect of water extracts of organic matter on plant pathogens. *Summa Phytopathologica,* 25: 330-35.

Naqvi S. A. M. H. (1986. Varietal screening of coriander against stem gall in relation to disease intensityand crop loss. *Ind. Jour. of Myco. and plant patho.,* 16(3): 270-276.

Narain, A. and Das, D. C. (1970). Toxin production during pathogenesis of *Colletotrichum capsici* causing anthracnose of chili. *Indian Phytopath.* 23: 484-490.

Narain, A. and Panigrahi, C. (1971). Efficacy of some fungicidal compounds to control *Colletotrichum capsici in vitro* and *in vivo. Indian Phytopath* 24: 593- 596.

Narasimha Rao, S., Ravindra, K. K. and Anandraj, M. (2012). Management of leaf spot of turmeric (*Curcuma longa* L.) incited by *Colletotrichum capsici* through fungicides. *Journal of Spices and Aromatic Crops.* Vol. 21(2): 151-154.

Narasimhudu, Y, and Balasubramanian, K. A. (2001b). Laboratory evaluation of fungicides against *Colletotrichum capsici* causing leaf spot of turmeric. *Pesrology.* 25: 31-32.

Narasimhudu, Y. and Balasubramanian, K. A. (2001a). Effect of fungicides on growth and vigour of turmeric. *Pesrology.* 25: 20-21.

Narula P. N and Joshi M. G (1963). Differential response of coriander strains to stem gall disease. *Science and Culture,* 29(4): 206.

Natarajan K (2002) Panchagavya: A manual, Other India Press, Mapusa, Goa, India, p. 33.

Ndoumou D. O, dzomoT. G. N and jocgoue P. F. D (1996). Changes in Carbohydrate, Amino Acid and PhenolContents in Cocoa Pods from Three Clones after Infection with *Phytophthoramegakarya* Bra. and Grif. *Ann. of Botany,* 77: 153-158.

Nduaguc, C., Ekefan, E. J. and Nwankiti, A. O. (2008). Effect of some crude plant extract on growth of *Colletotrichum capsici* (Synd.) Butler and Bisby, causal agent of pepper anthracnose, *Journal of Applied Bio-Sciences,* 6 (2): 184-190.

Neeraj and Vermashilpi (2010). Quantitative changes in sugar and phenolic contents of *Brassica* leaves induced by *Alternaria brassicae* infection. *International Journal of Plant Protection,* 3(1): 28-30.

Nene Y. L, Siddiqui I. A and Kharbanda P. D (1966). Control of stem gall of coriander by fungicide. *Mycopath. Mycol. Appl.,* 29(1-2): 142-144.

Nene, Y. L. and Thaplial, P. N. (1993). Evaluation of fungicides in; Fungicides in plant disease control (2011 new ed.) *Oxford, IBH Publishing Co.,* New Delhi, 331 p.

Nikam B. C, Kulkami, S. N and Pawar P. K (1959). Coriander its diseases. *Indian Forming,* Nov. 12-13.

Nishida H and SugiyamaJ (1994). Archiascomycetes: detection of a new lineage within the Ascomycota. *Mycoscience,* 35: 361-366.

Nishida H, Ando K, Ando Y, Hirata A and Sugiyama J. (1995). Mixiaosmundae: transfer from the Ascomycota to the Basidiomycota based on evidence from molecules and morphology. *Can. J. Bot.,* 73: 660-666.

Nyam M. A, Wonang D. L, Akueshi C. O (2009). Phytochemical screening and antimicrobialstudies on *Canarium schiveinfurthii* Linn ("Atili") fruits and oil. *Nigerian Journal of Botany,* 22 (2): 247-253.

Okigbo, R. N. and Nameka, I. A. (2005). Control of Yam tuber with leaf extracts of *Xylopia aethiopica* and *Zingiber officinale. Afr. J. Biotechnol.* 4(8): 804-807.

Okigbo, R. N. and Ogbonnaya, U. O. (2006). Antifungal effect of two tropical plants leaf extracts (*Ossimum gratinnimum* and *Aframomum meleguata*) on post harvest yam (*Diascortea* spp.) rot. *African Journal of Biotechnology*, 5: 727-731.

Ong CA, Vorghese G and Ting WP. (1986). The effect of chilli veinal mottle virus on yield of chilli (*Capsicum annuum* L.). *MARDI Res Bull*, 8: 74- 8.

Osbourn A. E (1996). Preformed antimicrobial compounds and plant defense against fungal attack. *Plant Cel*, 8: 1821-1831.

Osbourn A. E (2003). Saponins in cereals. *Phytochemistry*. 62: 1-4.

Palarpawar, M. Y. and Ghrude, V. R. (1989). Saprophytic colonization of leaf sheath of turmeric by *Colletotrichum capsici* and *Colletotrichum curcumae. Indian J of Mycol. PI Patho* 19: 202-203.

Palarpawar, M. Y. and Ghrude, V. R. (1997). Influence of different nitrogen sources on growth and sporulation of *Colletotrichum capsici* and (*Colletotrichum curcumae*) *J. of Mycol. And PI. Pathol.* 27: 227-228.

Pandey R. N and Dange S. R. S (1998). Diseases of coriander and fennel- a review. *Agricultural Review Karnal*, 19(2): 120-125.

Papadopoulou K, Melton R. E, Leggett M, Daniels M. J and Osbourn A. E (1999). Compromised disease resistance in saponin-deficient plants. *P. Natl. Acad. Sci. USA.*, 96 22): 12923-12928.

Papen H, Gabler A, Zumbusch E and Rennenberg H (2002). Chemo litho autotrophic nitrifiers in the phyllosphere of a Spruce ecosystem receiving high nitrogen input. *Curr. Microbiol.*, 44: 56-60.

Parihar Singh Pankaj (2012). Changes in metabolites of *Brassica juncea*(Indian mustard) during progressive infection of *AlternariaBrassicae. Nature and Science*, 10(3): 39-42.

Park, M. (1941). Report on the work of the mycological division, Ceylon Administration Reports. *Rep. Dir. Agric. Ceylon*, 1939: D-20-D-22.

Parmasto E and Parmasto I (1987). Variation of basidiospores in the Hymenomycetes and its significance to their taxonomy. *Bibl. Mycol*, 115: 1–168

Parveen, T. and Sharma, K. (2014). Management of "Soft Rot" of Ginger by Botanicals. *Int. J. of Pharm. Life Sci.* 5(4): 3478-3484.

Patel A. S and Patel H. R (1990). Biochemical changes brought out by the Infection of stunt nematode in Bidi tobacco. *Indian J. Mycol. Plant Pathol*, 20: 174-176.

Pathak M. L and Kumar a (2003). "Cow praising and importance of panchagavya as medicine, " *Sachitra Ayurveda*, 5): 56-59.

Pathak, M. I. and Kumar, A. (2003). Go-mutra destructive study. *Sachitra Ayurveda*. 7: 81-84.

Patil, A., Laddha, A., Lunge, A., Paikrao, H. P. and Mahure, S. (2012). *In vitro* antagonistic properties of selected *Trichoderma species* against tomato root rot causing *Pythium species*. *International J. Sci., Environ and Tech*, 1(4): 302–315.

Patil, M. J., Ukey, S. P. and Raut, B. T. (2001). Evaluation of fungisides and botanicals for the management of early blight (*Alternaria solani*) of tomato. *PKV Research Journal*, 25: 49-51.

Paul, B., K. Bala, L. Belbahri, G. Calmin, E. Sanchez- Hernandez and F. Lefort. (2006). A new species of *Pythium* with ornamented oogonia: morphology, taxonomy, ITS region of its rDNA, and its comparison with related species. *FEMS Microbiology Letters*, 254: 317-323.

Pavgi M. S and Mukhopadhyay A. N (1969a). Transmission of *Protomyces macrosporus* Unger, the incitant of stem gall of coriander. *Ann. Phyto. Patho. Soc. Japan*, 35(4): 265-270.

Pavgi M. S and Mukhopadhyay A. N (1969b). Viability of seed in coriander infected by *Protomyces macrosporus*Unger. *Ann. Phyto. Patho. Soc. Japan*, 35(4): 271-274.

Pavgi M. S and Mukhopadhyay A. N (1970). Cytology of chlamydospore germination in *Protomyces macrosporus* Unger. Faculty of Agriculture, Banaras University, India. *Cytologia*, 35: 359-367.

Pavgi M. S and Mukhopadhyay A. N (1972). Development of coriander fruit infected by *Protomyces macrosporus* Unger, 37: 619-627.

Pavgi M. S and Mukhopadhyay A. N (1973). Cytology of Chlamydospore Development in *Protomyces macrosporus* Unger. Faculty of Agriculture, Banaras Hindu University, India. *Cytologia*, 38: 467-477.

Paxton J. D (1981). Phytoalexins-a working redefinition. *Phytopathol Z*, 101: 106-109.

Payal Lodha and ParasharAnamika(2007). Phenolic estimation in *Foeniculumvulgare* infected with ramularia blight. *Annals of Plant Protection Sciences*, 15(2): 396-398.

Pinto S. S, Mann M. B, Campos F. S, Franco A. C, Germani J. C and Van Der SST (2010). Preliminary characterization of some *Streptomyces* species isolated from a composting process and their antimicrobial potential. *World Journal of Microbiology andBiotechnology*, 26(10): 1847-1856.

Plotnikova L. Z (1977). Dung infusion as a means for controlling powdery mildew of rose. *ZashchitaRastenii*, 2: 28.

Popta C. M. L (1899). BeitragzurKenntnis der Hemiasci Flora, 86: 1-46.

Poudyal, B. K. (2011). *Commercial Ginger Farming in Nepal, Today and Future*. Sajha Prakashan Press, Kathmandu, Nepal.

Prabhu M. J (2009). Panchagavya: low cost organic input for both crops and animals. In: Farmer's Notebook, *The Hindu Online Edition*, 4th June 2009.

Prasad B. K (1983). Free sugar and amino acid contents of coriander infected with stem gall disease. *Indian Phytopath.*, 36(1): 150-151.

Prasad B. K, Singh R. N, Shankar U and Krishore A (1989a). Alteration in in the respiration of coriander infected with *Protomyces macrosporus* Unger causing stem gall. *Indian Phytopath.*, 42(4): 534-537.

Prasad B. K, Singh R. N and Narayan, N (1989b). Biochemical changes in nitrogen and carbohydrate in coriander infected with *Protomyces macrosporus* Unger. *Indian Phytopath.*, 42(3): 426-430.

Prashith Kekuda, T. R., Vivek, M. N., Manasa, N., Kambar, Y., Noornawaz, A. S. and Raghvendra, H. L. (2014). Antifungal effect of cow urine extracts of selected plants against *Collrtotrichum capsici* isolated from anthracnose of chili. *International Journal of Sciences.* 7 (3): 142-143.

Price K. R, Johnson, I. T and Fenwick, G. R (1987). *CRC Crit. Rev. Food Sci. Nutr.*, 26: 27-133.

Pring, R. J., Nash, C., Zakaria, M. and Bailey, J. A. (1995). Infection process and host range of *Colletotrichum capsici. Physiol. Mol Pl. Parhol.* 46: 137-152.

Pruthi J. S (1976). Spices and condiments. *4th Ed.*, National Book Trust, India, New Delhi, pp. 1-286.

Pruthi, J. S. (1998). Major spices of India Crop management, post-harvest technology. Publ. Publication and Information Div., ICAR, New Delhi.

Pureifull DE, Zitter TA and Hiebert E. (1975). Morphology, host range, and serological relationships of pepper mottle virus.

Purseglove, J. W. (1968). Turmeric. In Spices. Vol. 2. Longman Inc. New York, pp. 532-579.

Rai, Bimla, M. M. Jha, K. L. Ojha and S. B. Sharma (1998). Effect of fungicidal seed treatment on Rhizome rot disease of ginger. *Nat. Sem. Rec. Dev. Spi. Tech.* B. A. C. April, 20-21, 1998 (pp. 43-44).

Raj Abhishek, Kumar Manoj, Toppo Pratap Jhariya (2014). Cow dung for ecofriendly and Sustainable productive farming. *Environmental Science,* 3: 2277-8179.

Raja J and Kurucheve V (1999). Fungicidal activity of buffalo (*Babulusbubalis*) urine: a new record. *Madras Agricultural Journal* 86(10/12): 614-616.

Rajak, C. R (1983). Nutritional requirements of two species of *Colletotrichum. Acro Boranico Indico.* 11: 155-160.

Rajan, K. M. and Agnihotri, V. P. (1989). *Pythium* induced rhizome rot of ginger-problems and progress. In: Agnihotri V P,

Rakesh K. N, Dileep N, Syed Junaid, PrashithKekuda T. R, Vinayaka K. S and Noor Nawaz A. S (2013). Inhibitory Effect of Cow Urine Extracts of Selected Plants against Pathogens Causing Rhizome Rot of Ginger. *Science, Technology and Arts Research Journal,* 2(2): 92-96.

Ram, D. and Patil, R. K. (2014). Evaluation of fungicides, antagonists, phytoextracts and cow urine against *Fusarium pallidoroseum* causing fruit rot of tomato. *J. Pl. Dis. Sci.,* 9(1): 100-104.

Ram, P., Mathur, K. and Lodha, B. C. (1999). Integrated management of rhizome rot of ginger involving biocontrol agents and fungicides. *J. Myco. and Pl. Patho.,* 29(3): 416-420.

Ramakrishnan, T. S. (1949). The occurrence of *Phytium vaxans* de Bary in South India. *Indian Phytopath,* 2: 27-30.

Ramakrishnan, T. S. (1954). Leaf spot disease of turmeric (*Curcumu irngu* L) caused by *Colletotrichum capsici* (Syd.) Butler and Bisby. *Indran Phytopath.* 7: 111-117.

Ramkumar, Suoreche, R., Prabhakar, S. and Pandurangan, M. (2012). Role of antagonistic microbe *Psuedomonas florescense* on *Colletotrichum capsici* infecting Curcuma longa. *J Plant Pathol.* Microb 3: 146. doi: 10. 4172/2157-7471. 1000146.

Ramteke, P. K. and Kamble, S. S. (2011). Physiological studies in *Fusarium solani* causing rhizome rot of ginger (*Zingiber officinale* Rosc.). *The Bioscan.* 6(2): 195-197.

Rangaswami, G. and Mahadevan, A (2001). Diseases of crop plants of India. IV Edition, (Eds.) Prentice hall of India (P.) Ltd., New Delhi. pp. 453-455.

Rao C. P (1972). Anatomical studies on abnormal growth caused by *Protomyces macrosporus* on *Coriandrum sativum* L. *Indian Phytopath.,* 25(4): 483-496.

Rao, C. H. and Narayana, Y. D. (2005). *In vitro* evaluation of fungicides, plant extracts and bio-control agents against *C. dematum* (Pers. Ex. Fr.) Grove the causal organism of chickpea (*Cicer arietenum* L.) blight. In national symposium on crop disease management in dry land Agril. and 57th Annual meeting IPS, Jan. 12-14, 2005, MAU, Parbhani.

Ravishanker, Kumar, S., Chatterjee, A., Baranwal, D. K. and Solankey, S. S. (2014). Genetic variability for yield and quality traits in ginger (*Zingiber officinale* Roscoe). *The Bioscan,* 8(4): 1383-1386.

Reddy M. S and Kramer C. L (1975). A taxonomic revision of the Protomycetales. *Mycotaxon,* 3: 1–50.

Reddy, S. G., Dakshinamurthy, V. and Sarma, S. S. (1963). Note on varietal resistance against leaf spot disease in turmeric. *Andhra Agric J.* 10: 146-148.

Rezk M. A, Ibrahim, I. K. A and Ibrahim A. A. M (1976). Effect of root-knot nematodes on phenolics contents of barley and wheat. *Nematol. Medit.* 15: 259-263.

Ricker, A. J. and Ricker, R. S. (1936). Introduction to research on plant diseases. John, S Swift Co., 4. 4, USA.

Roberts, R. G. and Snow, J. P. (1990). Morphological and pathological studies of *Colletotrichum capsici* and *C. indicum. Mycologia.* 82: 82-90.

Rohit R. P, Kavindra R. J and Chintan K. M, (2011). Image morphological operation based quality analysis of coriander Seed (*Coriandrum sativum* L.) Emerging Trends in Networks and Computer Communications (ETNCC). *International Conference,* 22-24: 482-486.

Rojasara, Y. M., Rajani, M. D., and Dhruj, I. U. (2014). Management of groundnut rust through host plant resistance and fungicides. *J. Pl. Dis. Sci.,* 9(1): 38-42.

Romero, E., Zurera, C., Porras, M., Barrau, C. and Romero, F. (2007). Biofumigant effect of *Brassica* species in the growth of *Phytophthora* spp. *in vitro.* 15th International Reinhardsbrunn Symposium: Modern Fungicides and Antifungal Compounds. Friedrichroda, Germany 6-10 May.

Ross, I. A. (1999). Medicinal Plants of the World. Chemical constituents, traditional and modern medicinal uses. Humana Press.

Sagar, S. D. (2006). Investigations on the etiology, epidemiology and integrated management of rhizome rot complex of ginger and turmeric. Ph. D. Thesis, Department of Plant Pathology, University of Agricultural Sciences, Dharwad.

Sagar, S. D., Srikant kulkarni, Hedge, Y. R., Rao, M. S. L. (2007). Management of rhizome rot of ginger by bio agents. *Annals of Agriculture Bio Research,* 12(1): 57-59.

Sahib N. G, Anwar F, Gilani A. H, Hamid A. A, Saari A and Alkharfy K. M (2012). Coriander (*Coriandrum sativum* L.): A potential source of high-value components for functional foods and nutraceuticals-*A Review. J. Phytother. Res.,* 27(9): 1002/ ptr. 4897.

Sahni, S. S., Chabal, D. S. and Nirmaljit Singh. (1974). *In-vitro* production of toxic metabolite by different species of *Colletotrichum. lndian J Mycol. PI Parhol.* 4: 222-223.

Sangamaheswara Sarma and Krishna Murthy, D. (1962). Varietal resistance against leaf spot diseases of turmeric. *Andhra Agric J.* 9: 6165.

Sangeetha V, Thevanathan R (2010a). Effect of TMV strains in a local lesion host. Journal of panchagavya on nitrate assimilation by experimental Ecobiol., 31: 28-131. plants. *The Journal of American Science,* 6: 80-86.

Sanger, R. B. S., Katawale, T. R., Sarof, R. K. and Parihar, M. S. (1988). *Farm Sci. Jour.* 3: 69-71.

Sanjay Kumar, Rajesh Kumar, Sanjeet Kumar, Major Singh, A. B. Rai And Mathura Rai (2009). Reaction of pepper leaf curl virus field resistance of chilli (*Capsicum annuum* L.) genotypes under challenged condition. *Veg. Sci.* 36(2) : 230-232.

Sanjutha S, Subramanian S, Rani I. and Maheswari (2008). Integrated Nutrient Management in *Andrographispaniculata. Research Journal of Agriculture and Biological Sciences,* 4(2): 141-145.

Sarma, Y. R. (1994). Rhizome rot disease of ginger and turmeric. In: Chadha K L and Rethinam P (Eds.) Advances in Horticulture, Vol. 10. Plantation and Spices Crops, Part-2. (pp. 1113-1138). Malhotra Publication House, New Delhi.

Sarwar, M., Kirkegaard, J. A., Wong, P. T. W., and Desmarchelier, J. M. (1998). Bio-fumigation potential of brassicas. In vitro toxicity of isothiocynates to soil borne-fungal pathogens. *Plant and Soil,* 201: 103-112.

Saunders, W. H. M. (1982). Effect of cow urine and its major constituents on pasture properties. *NZ J. Agricul. Res.* 25 (1): 61-68.

Saxena M, Saxena J, Nema R, Singh D and Gupta A (2013). Phytochemistry of medicinal plants. *J. Pharm. Phytochem.,* 1(6): 168-82.

Saxena R. P, Dixit, J, Pandey U. P and Singh V. K (2002). Germplasm screening for stem gall disease of Coriander. Paper presented in National Symposium on integrated management of plant disease of Mid-Eastern India with a cropping system perspective, held at NDUAT, Faizabad, Dec. 5-7-2002.

ScarpariL. M, Meinhardt L. W, Mazzafera, PomellaP. A. Schiavinato A. M, CascardoJ. C. M and PereiraG. A. G (2005). Biochemical changes during the development ofwitches' broom: the most important disease of cocoa in Brazil caused by *Crinipelli sperniciosa*. *Journal of Experimental Botany*, 56(413): 865-877.

Schoemaker, I. A. S. Ghini, R. (2001). Bio-fumigation for *Pythium spp.* control (Portguese). *Summa Phytopathologica*; 27(3): 308-312.

Schönbeck F and Schlösser E (1976). Preformed substances as potential protectants. In: Heitefuss R, Williams P H, editors. Physiological plant pathology. Berlin, Germany: *SpringerVerlag* KG, 653-678.

Schrandt, J. K., Davis, R. M. and Nunez, J. J. (1994). Host range and influence of nutrition, temperature and pH on growth of *Pythium violae* from carrot. *Plant Dis.* 4(78): 335-338.

Senanayake, D. M. J. B., Verma A., and Mandal, B. (2012). Virus- Vector Relationship, Host Range, Detection and sequence comparison of chilli leaf curl associated with an epidemic of leaf curl disease of chilli in Jodhpur, India. *J. Phytopathol* 160: 146-155.

Senapati, G. R. and Sugata Ghose (2005). *Indian Phytopatho.*, 58: 437-439.

Sengupta, D. K., Maity, T. K. and Dasgupta, B. (2009). Effect of mulching on ginger (*Zingiber officinale* Rosc.) in the hilly region of Darjeeling district. *Journal of Crop and Weed*, 5(1): 203-205.

Shah H, Khalid S and Ahmad I. (2001). Prevalence and distribution of four pepper viruses in Sindh, Punjab and North West Frontier Province. *On Line J Biol Sci*, 1: 214-217.

Shahare, K. C. and Asthana, R. P. (1962). Rhizome rot of ginger and its control, *Indian Phytopath.*, 15 (1): 78 (RAM) 42(1): 41, 1963.

Shailaja B, Mishra Ipsita, Gampala Srihima, Singh Vikram Jeet and Swathi K (2014). Panchagavya- an eco-friendly insecticide and organic growth Promoter of plants. *International Journal of Advanced Research*, 2: (11) 22-26.

Shakuntala N. M, Vasudevan S. N, Patil S. B, Doddagoudar S. R, Macha RCMSI Vijay and Kumar A. G (2012). Organic bio priming on seed vigour inducing enzyme in paddy - An alternative to inorganics. *Ecoscan*, 1: 251-257.

Shalika (2009). Non-chemical management of stalk rot of cauliflower caused by *Sclerotinia sclerotiorum* (Lib.) de Bary. M.Sc. Thesis, Department of Plant Pathology, CSK Himachal Pradesh KrishiVishvavidyalaya, Palampur, India, p. 48.

Sharma A and Deshpande V. K (2006). Effect of pre-soaking of pigeon pea seeds with organics on seed quality. *Karnataka Journal of Agricultural Sciences*, 19: 396-399.

Sharma O. P. and Singh J (1985). Reaction of different genotypes of pepper to cucumber mosaic and tobacco mosaic viruses. *Caps. Newsl.* 4: 47.

Sharma, B. R., Dutta, S., Roy, S., Debnath A. and Roy, M. D. (2010). The effect of soil physico-chemical properties on rhizome rot and wilt disease complex incidence of ginger under hill agro-climatic region of West Bengal. *Plant Pathology J.*, 26(2): 198-202.

Shepherd RJ and Purcifull DE (1971). Tobacco etch virus CMI/AAB, In: Descriptions of plant viruses, pp: 55. Taufik M, Astuti AP, and HidayatSH. 2005. Survey of *Cucumber mosaic virus* and *Chilli veinal mottle virus* on chilli pepper and screening of resistance of chillipepper genotypes (in Bahasa Indonesia, English abstract). *J. Agricult.*, 16: 146-152.

Shivapuri, A., Sharma, O. P. and Jhamaria, S. L. (1997). Fungitoxic properties of plant extracts against pathogenic fungi. *Journal of Mycology and Plant Pathology;* 27: 29-31.

Siddiquie, M. R., Dharam Sigh and Ashok Gaur (1977). Prevalence of chilli anthracnose hgus on seeds and its effective control. *Seed Res.* 5: 67-72.

Sindhan G. S, Parashar R. D and Hooda, I (1996). Relationship between biochemical parameters and flag smut resistance in wheat. *Indian J. Mycol. Plant Pathol.*, 26: 291-293.

Singh G and Bedi D. S (1976). Phenolic and sugar constituents of gram cultivarsresistant and susceptible to *Operculella padwickii. Indian Phytopath.*, 29: 191-192.

Singh G. R, Chaure N. K and Parihar S. S (2001). Organic Farming for sustainable agriculture *Indian Farmer*, 52: 12-17.

Singh H. B, Singh A, Tripathi A, Rai S. K, Katiyar R. S, Johri J. K, and Singh S. P. (2003). Evaluation of Indian coriander accessions for resistance against stem gall disease. *Genetic Resources and Crop Evolution*, 50: 339-343.

Singh K, Vashistha RN, Pandita ML and Batra, BR (1976). Effect of mulching on growth and yield of cucurbits under rainfed conditions. *Haryana J. Hort. Sci.*, 5(1 and 2): 87-91.

Singh P and Thapliyal P. N (1987). Effect of some soil factors on disease controlling potential of Carbendazim. *Indian Phytopath.*, 40(2): 204-208.

Singh S. P, Gupta J. S and Sharma A. K (1984). Disease appraisal and crop loss estimates in coriander attacked by *Protomyces macrosporus* Unger in Uttar Pradesh. *Geobios*, 11(6): 276-278.

Singh SJ (1973). Reaction of chilli Cultivars (*Capsicum* sp.) to mosaic and leaf curl viruses in field conditions. *Indian J. Hort.* 30(1-2): 444-447.

Singh, A. K. and Edison, S. (2003). Eco Friendly management of leaf spot of turmeric under partial shade. *Indian Phytopathology.* 56: 4, 479-480. S

Singh, A. K., Basandari, A. and Sharma, B. K. (2003). Fungicidal management of leaf spot of turmeric. *Indian Phytopathology.* 56 (1): 119-120.

Singh, A. K., Singh, R. K., Singh, A. K., Singh, V. K., Rawat, S. S., Mehta, K. S., Kumar, A., Gupta, M. K. and Thakur, S. (2014). Bio-mulching for ginger crop management: Traditional ecological knowledge led adoption under rainfed agro-ecosystems. *Indian J. Traditional Knowledge*, 13(1) 111-122.

Singh, H. K., Shakywar, R. C., Shyam Singh, Singh, A. K. (2012). Evaluation of comparative efficacy of native isolate of *Trichoderma viride* against rhizome rots disease of ginger. *J. Pl. Dis. Sci.*, 7(1): 22-26.

Singh, R. S. (1963). Plant disease Oxfords IBH Publishing Co., New Delhi-1.

Singh, R. S. (2002). Introduction to Principles of Plant Pathology. Fourth Edition, Oxford and IBH Publishing Co. Pvt. Ltd., New Delhi. pp. 279-292.

Singh, R. S. (2006). Plant Disease. 8th Edition Oxford and IBH Publishing Co. Ltd. New Delhi, 167-168.

Singh, S. Gupta, B. R., Pande, S. P. and Tripathi, D. P. (1990). Response of some chilli lines to mosaic and leaf curl diseases, *Indian J. Virol.*, 6(1-2): 113-115.

Singh, S. J., Sastry, K. S. and Sastry, K. S. M. (1979). Combating leaf curl virus in chilli. *Ind. Hort.*, 24: 9-10.

Singh, S. J., Sastry, K. S. and Sastry, K. S. M. (1979). Efficacy of different insecticides and oil in the control of leaf curl virus in chilli. *Zeit. Pflanzens.*, 86 (5): 253-256.

Singh, S. K., Rai, B., Kumar, B. (2004). Evaluation of different fungicides in controlling the rhizome rot of ginger under storage and field conditions. *Annals of Agri Bio Research*, 9(1): 63-65.

Singh, S. P.. Production Technology og Vegetable Crops. Agricultural Communication Research Centre, Karnal, Haryana

Singh, S. P. (1999). Biological Control in India. *Ind J PI Pror*. 126-138

Singh Amitabh, Kumar Jaiswal. Ritesh, Maurya Sudarshan, and Singh Udai Pratap (2013). Analysis of phenolic and indole acetic acids in *Meloidogyne graminicola* infected rice plants (*Oryza sativa* L.). *International Journal of Advanced Research*, 1(6) 71-76.

Sinha R. K, Valani D, Chauhan K and Agarwal S. (2010). Embarking on a second green revolution for sustainable Agriculture by vermin culture biotechnology using earthworms: Reviving the dreams of Sir Charles Darwin. *Journal of Agricultural Biotechnology and Sustainable Development*, 2: 113-128.

Sinha, B., and Trafdar, J. (2014). Epidemiological study of SPVD transmission in West Bengal. *Bioscan*. 9(3): 1237-1242.

Sivan A., Y. Elad and I. Chet. (1984). Biological control effects of a new isolate of *Trichoderma harzianum* on *Pythium aphanidermatum*. *Phytopathology*, 74: 498-501.

Snapp, S. S., O'Neil, K., Kirk, W. W., Bird, G., Quakenbush, B. (2004). Improving short potato-based rotations through organic amendments. Michigan 2004 Potato Research Report. Michigan State University Agricultural Experiment Station, in cooperation with The Michigan Potato Industry Commission 36: 22-27.

Somasundaram, E and Amanullah M. M (2007). Panchagavya on growth and productivity of crops: *A review. Green Farming,* 1: 22-26.

Sparg S. G, Light M. E and Staden J (2004). Biological activities and distribution of plant saponins. *J Ethnopharmacol*, 94: 219-243.

Sreenivasa M. N and Naik N (2011). Impact of bacteria isolated from organic liquidmanures on seed germination and seedling vigour index of wheat and soybean. *Organic Farming Newsletter*, 7: 3-5.

Srikant Kulkarni (2011). Etiology, epidemiology and integrated management of rhizome rot of ginger in Karnataka. *Indian Phytopath.*, 64(2): 113-119.

Srimal, R. C. (1997). Turmeric: A brief review of medicinal properties. *Fitoperapia* 68(6): 483-493.

Srimathi P, Mariappan N., SundaramoorthyL and Paramathma M (2013). Efficacy of Panchagavya on seed invigoration of biofuelCrops. *Academic Journals.* 8(41): 2031-2037.

Srivastava H. C (1955-b). Studies on tumour disease of *Coriandrum sativum* L. caused by *Protomyces macrosporus* Unger. *Proc. 42nd Ind. Sci. Cong.*, Part III: 218.

Srivastava U, Rai R. A and Agrawal J. M (1971b). Powdery mildew of coriander and its control. *Indian Phytopath.*, 24(3): 437-440.

Srivastava, V. P. and Gupta, J. H. (1977). Fungicidal control of turmeric leaf spot incited by *Taphrina maculans. Ind J Mycol. PI Pathol.* 7: 76-77.

Stanghellini, M. E. and Russell, J. D. (1973). Germination *In vitro* of *Pythium aphanidermatum* oospores. *Phytopathology*, 63: 133-137.

Stansly, P. A., Schuster, D. J. and Liu, T. X. (1997). Apparent parasitism of Bemisia argetifolii (Homoptera: Aleyrodidae) by Aphelinidae (Hymenoptera) on vegetable crops and associated weeds in South Florida. *Biological control*, 9: 49-57.

Stephanie, G. Harvey and Carl E. Sams. University of Tennessee (2001). Brassica bio-fumigation increases marketable tomato yield, Department of Plant Sciences and Landscape Systems, Knoxville, TN 37901.

Stirling, G. R., Turaganivalu, U., Stirling, A. M., Lomavatu, M. F. and Smith, K. M. (2009). Rhizome rot of ginger (*Zingiber officinale*) caused by *Pythiu m myriotylum* in Fiji and Australia. *Australasian Plant Pathology*, 38, 453–460.

Subhashini, D. V. (2013). Bio-efficacy of plant extracts against major pathogens affecting tobacco nurseries, *Annal Plant Protection Science*: 20(2): 414-416.

Subrahmaniyan, K., Zhou, W. and Veeramani, P. (2011). Weed control through degradable, herbicidal plastics and organic mulches and their effect on crop growth and yield of winter rapeseed (*Brassica napus*). *Indian J. Agril. Sci.*, 81(4): 348-352.

Subrahmaniyan, L. S. (1919). A *Pythium* disease of ginger. *Madras Agric. J.*, 38: 225-226.

Sugha S. K (2005). Antifungal potential of panchgavya. *Plant Disease Research* 20: 156-158.

Suleiman, M. N., Emua, S. A. and Ayodele, S. M. (2011). Growth and physiological studies on root rots fungus of cowpea. *European Journal of Experimental Biology*, 1(4): 181-188.

Suprapta, D. N. and Khalimi, K. (2009). Efficacy of plant extracts formulations to suppressed stem rot disease on Vanilla seedlings. *J. ISSAAS Vol.* 15 No. 2: 34-41.

Swaminathan C, Swaminathan V and Vijay Lakshmi K (2007). Panchgavya boon toorganic farming. *International Book Distributing Co. Lucknow*, p. 332.

Swamy, B. S. and Kulkarni, S. (2003). *In vitro* evaluation of fungicides and botanicals against *Colletotrichum capsici* (Syd) causing leaf spot of turmeric. *Indian Phytopath.* 56 (3), 339.

Sydow, H. (1913). Beitrage Zur Kenntnis der Pilzflora des sudlichen Osrindiens I *Ann Mycol.* 11: 329-330.

Tambe V. D, Bhambar R. S (2014). "Estimation of Total Phenol, Tannin, Alkaloid and Flavonoid in *Hibiscus Tiliaceus* Linn. Wood Extracts" *RRJPP, 2(4)*: 41-47.

Tarafdar, J. and Saha, N. (2007). Correlation study on population dynamics of ginger soft rot inciting pathogens under different organic amendments, disease incidence and its survival in Darjeeling hill soils. Proceedings of the 13th ISTRC Symposium, pp. 165-169.

Tayal K. S, Kumar S, Geol A. K and Maheshwari D. K (1981). Role of IAA, IAA oxidase o-dihydroxy phenols, poly phenol oxidase and peroxidase in stem gall disease of *Coriandrum sativum* L. *Current Science*, 50(17): 785-786.

Tewari, V. P., Ramanujam, S. (1974). Grow Pusa Jwala, a disease resistant high yielding chilli. *Indian Farming* 24, 20.

Thakur S (2010). Studies on some plant growth promoting traits in bacteria isolated from vermiwash. M Sc Thesis. Department of Microbiology, College of Agriculture, CSK HPKV, Palampur, India, p 86.

Thamburaj, S. (1991). Research on spice crops at TNGDNAU. *Spice India.* 4: 17-21.

Tharmaraj K., Ganesh P., Suresh Kumar R., Anandan A. and Kolanjinathan K. (2011). "A critical review on Panchagavya—a boon plant growth," *International Journal of Pharmaceutical and Biological Archive*, 2(6): 1611–1614.

Thind, K. S. and Randhawa, H. S. (1957). Studies on the nutrition of (*Colletotrichum capsici* (Syd.) Butler and Bisby), the incitant of die-back of chilli. *Curr Sci* 26: 17-18.

Thirumalachar M. J and Pavgi M. S (1950). Notes on spore germination and mounting techniques. *Indian Phytopath.*, 3: 177-178.

Thirumalachar M. J, Pavgi M. S, and Singh R. A (1969). *In vitro* activity of 3 antifungal antibiotics against *Protomyces macrosporus* Unger, *Tapharina maculans* and *Corticum sasakii. Hindustan Anti. Biot. Bull.*, 11(3): 189-190.

Thirumalachar M. J and Narasimhan M. J (1953). Notes on some mycological methods. *Mycologia*, 45: 461-468.

Thomas, K. M. (1938). Detailed administration Report of the Govt. Mycologist, Madras, 21 pp.

Thomas, K. M. (1940). Detailed administration Report of the Govt. Mycologist, Madras, 30 pp.

Thomas, K. M. (1941). Detailed administration Report of the Govt. Mycologist, Madras, 18 pp.

Tiwary R. S and Agrawal A (2004). Production technology of spices. *1st ed. International Book Distribution Co.* Chaman Studio Building, 2nd floor, Charbagh, Lucknow, 2260004 U. P., India, pp. 254-271.

Toncer O, Tansi S and Kizil S (1998). The effect of different sowing time on essential oil of coriander in GAP region. *Anadolu.*, 8(2): 101-105.

Tripathi A. K (2001). Evaluation of coriander cultivars for resistant against stem gall disease and yield potncial. *Crop Research Hisar*, 22(3): 485-488.

Tripathi A. K (2003). Effect of sowing dates and weather parameters on stem gall disease of coriander. *Indian Phytopath.*, 56(2): 191-193.

Tripathi A. K and Bartaria A. M (2002). Chemical control of stem gall disease (*Protomyces macrosporus*) of coriander and its economics. *Plant Disease Research*, 17(1): 119-120.

Tripathi A. K, Pandya R. K, Tiwari R. J and Tripathi M. L (2002). Response of stem gall disease (*Protomyces macrosporus*) and yield of coriander (*Coriandrum sativum*) to soil pH. *Indian J. Agri. Sci.*, 11: 49-50.

Tubki K (1957). Biological and cultural studies of three species of *Protomyces. mycologia*, 49: 44-54.

Tuite, J. (1969). Plant pathological methods fungi and bacteria, Burgess Publishing Company, Minneapolis Min. 55

Turner E. M (1953). *J. Exp. Bot.*, 4: 264- 271.

Usman, N. M., Balakrishnan, P. and Sarma, Y. R. (1996). Biocontrol of rhizome rot of ginger. *Journal of Plantation Crops*, 24: 184-191.

Van der Plaats-Niterink, K. J. (1981). Monograph of the genus *Pythium*. Stud. Mycol. 21: 1-24.

Van Os, G. and L. Lazzeri, (2006). Control of *Rhizoctonia solani* in lily by biofumigation with *Brassica* seed meal. Proceedings of the 2nd International Biofumigation Symposium, June 25-29, University of Idaho, Moscow.

Vasudeva, R. S. (1954). Report of the division of mycology and Plant Pathology. *Sci. Rept. Agric Res* 79-89.

Vasumathi R (2001). Influence of organic manures, bio fertilizers and plant density on growth, yield and alkaloid content of Bhumyamalaki (*PhyllanthusamarusSchum.* And Thonn.) M.Sc. (Hort.) Thesis TNAU, Coimbatore-3. India.

Verma, A. K., Basu, D., Nath, P. S., Das, S. and Ghatak, S. S. and Mukhopaghyay, S. (1989). Relationship between the population of whitefly, *Bemisia tabaci* Genn. (Homoptera: Aleyrodidae) and the incidence of tomato leaf curl virus disease. *Ind. J. Mycol. Res.* 27 (1): 49-52.

Verma, R. K., Singh, Y., Soni, K. K. and Jamalluddin (2005). Solarization of forest nursery soil for elimination of root pathogens and weeds. *Indian J. Trop. Biodiv.*, 13: 81-86.

Vikas and Ratnoo, R. S. (2011). Evalution of fungicides and phytoextract against powdery mildew (*Erysiphe polygoni* de.) of fenugreek. Journal of Plant Disease Scinces. Vol-6 issue 1. Pp-71-73.

Vrang, J., Rasochova, M., Ficker, A. and Dobias, K. (1990). Inoculation of potatoes with micro-organisms under field conditions. 1. Effect of plant growth, yield and physiological properties of tubers in potato and sugerbeet. *Folia Microbiologica,* 35: 326-335.

Wate, S. T., Dhanjode, D. P., Duragkar, M. J. and Tajne, M. R. (2011). Antioxidant potential of cow urine and its fractions: A comparative *study: Inter J. Uni Pharm Life Sciences.* 1(1): 146-154.

Watve, Y. G., Diwakar, M. P., Kadam, J. J. and Sawant, U. K. (2009). Studies on effect of different fungicides on *Colletotrichum gloesporioides* penz. causing leaf spot disease of Jatropha. *J. Pl. Dis. Sci.,* 4(1): 120-121.

Watve, Y. G., Diwakar, M. P., Sawant, U. K. and Kadam, J. J. (2009). Studies on effect of different fungicides and Phyto-extract on *Colletotrichum gloesporiodes* Penz. Causing leaf spot of Jatropha. *J. Pt. Dis. Sci.* 4(1): 95-98.

Wei Y. D, Shen W. Y, Dauk M, Wang F, Selvaraj G and Zou J. T. (2004) Targeted gene disruption of glycerol-3-phosphate dehydrogenase in *Colletotrichum gloeosporioides* reveals evidence that glycerol is a significant transferred nutrient from host plant to fungal pathogen. *Journal of Biological Chemistry,* 279: 429-435.

Xu H. L (2001). Effects of a microbial inoculant and organic fertilizers on the growth, photosynthesis and yield of sweet corn. *J. Crop Prod.,* 3: 183-214.

Xu H. L (2001). Effects of a microbial inoculants and organic fertilizers on the growth, photosynthesis and yield attributes and economics of rice (*Oryza sativa*). *Crop Res.,* 31: 1-5.

Yadav B. K and Lourdraj C. A (2006). Effect of organic manures and Panchagavya spray on yield attributes and economics of rice (*Oryza sativa*). *Crop Res.,* 31: 1-5.

Zaman, M. M., Masum, A. S. M. H., Ahmed, M. A., Salam, M. A. and Rahman, M. H. (2002). Effect of tillage and mulch on the growth and yield of ginger in the Hilly area. *J. Biological Sci.,* 2(2): 121-123.

Zhuang Lu Yang (1992). Extraction and determination of flavonoid in ginkgo *Chinese Herbal Medicine,* 23: 122–124.

Zitter TA. (1972). Pepper virus strain identification in southern florida. *Phytopathology,* 62: 800.

Index

P

R

S

F

G

H

I

K

L

M

N

O

P

R

S

p. 9

p. 10

p. 22

p. 22

p. 25

p. 26

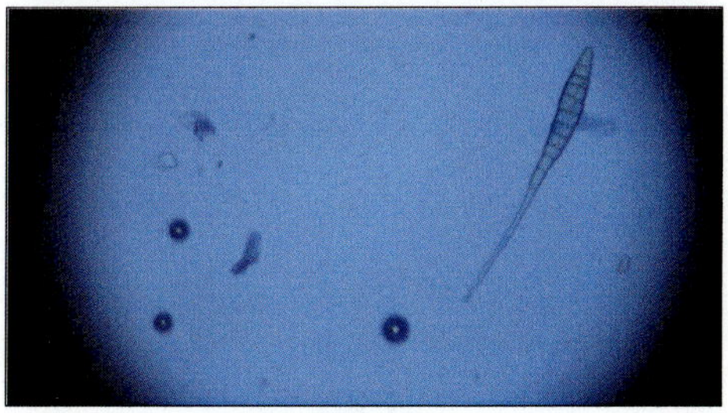

p. 26

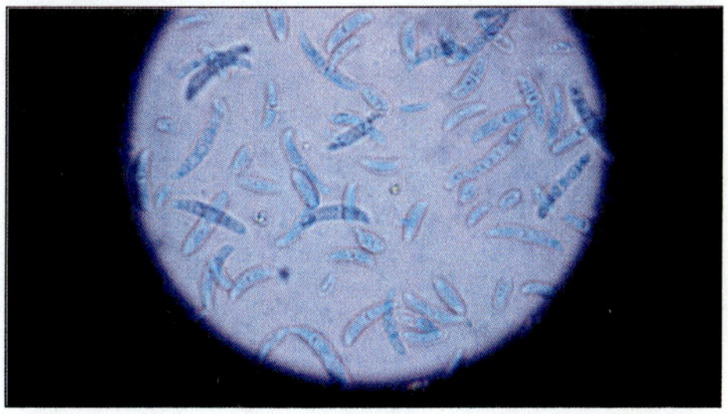

p. 27

p. 28

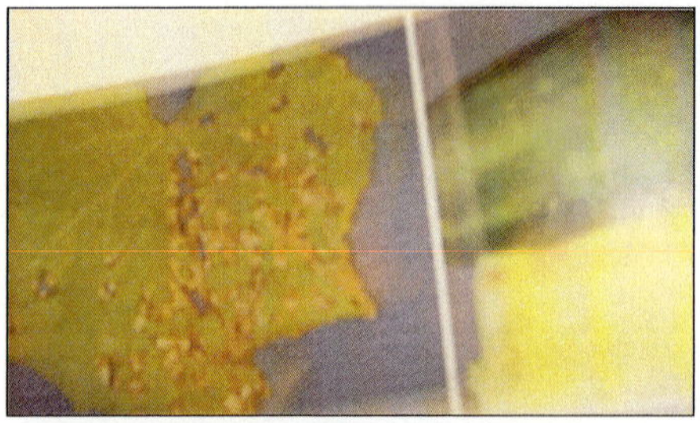

p. 31

p. 34

p. 43

p. 44

p. 45

p. 46

p. 46

p. 47

p. 48

p. 49

p. 50

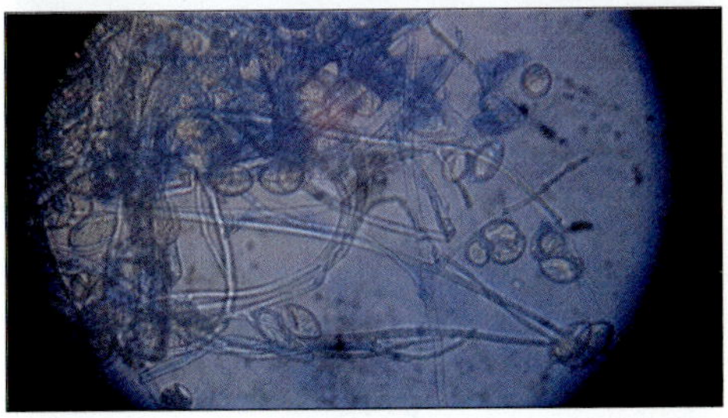

p. 53

p. 54

p. 54

p. 57

p. 58

p. 59

p. 62

p. 64

p. 67

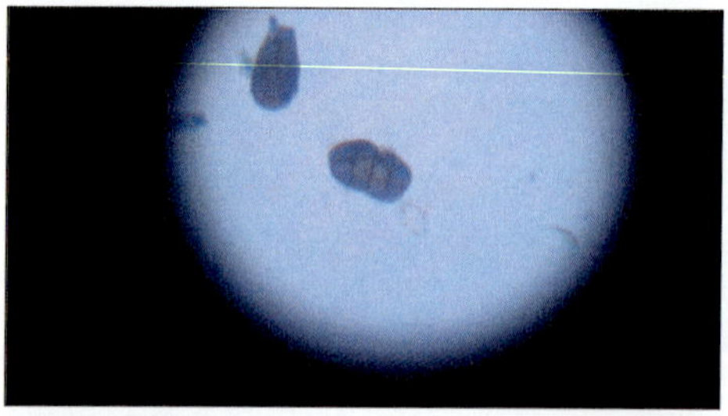

p. 73

p. 78

p. 87

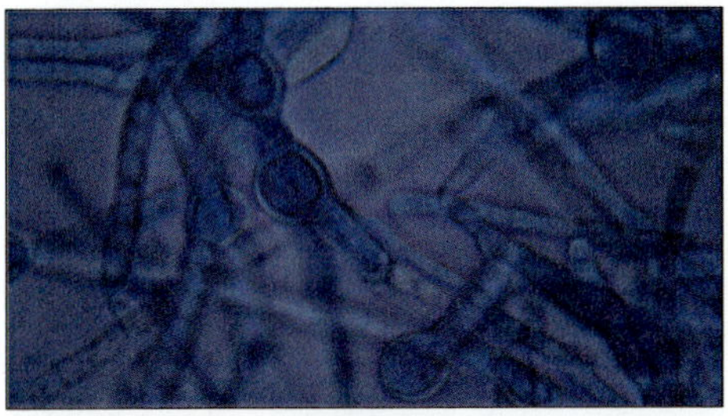

p. 88

p. 89

p. 91

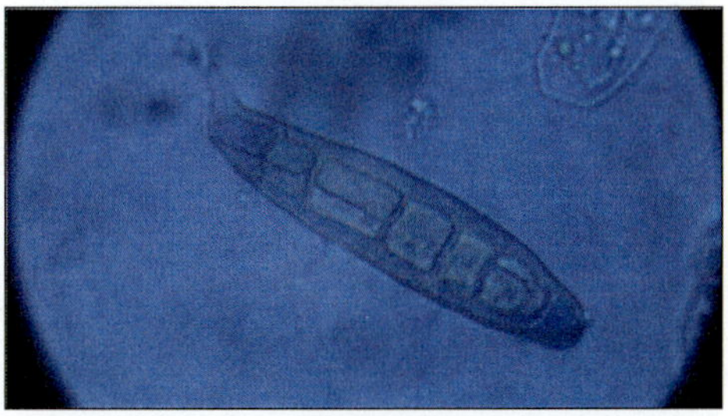

p. 91

p. 92

p. 92

p. 93

p. 95

p. 95

p. 100

p. 100

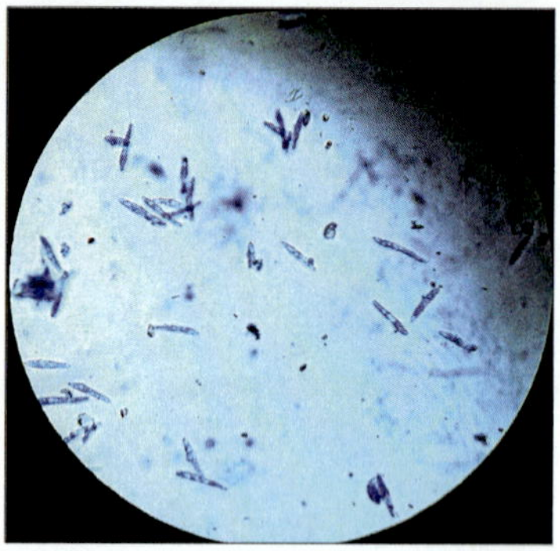

p. 101

p. 103

p. 107

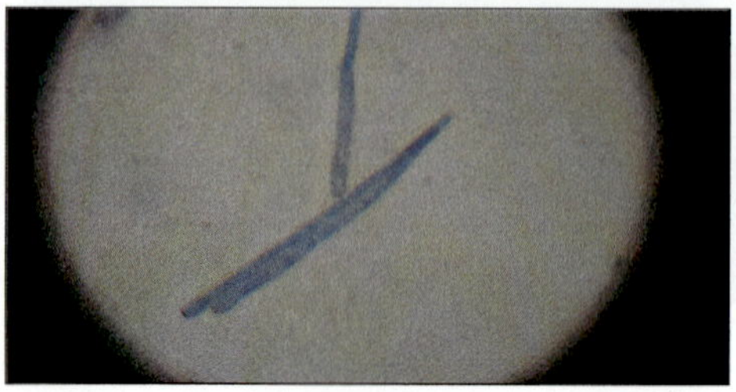

p. 107

p. 108

p. 109

p. 109

p. 113

p. 119

p. 120

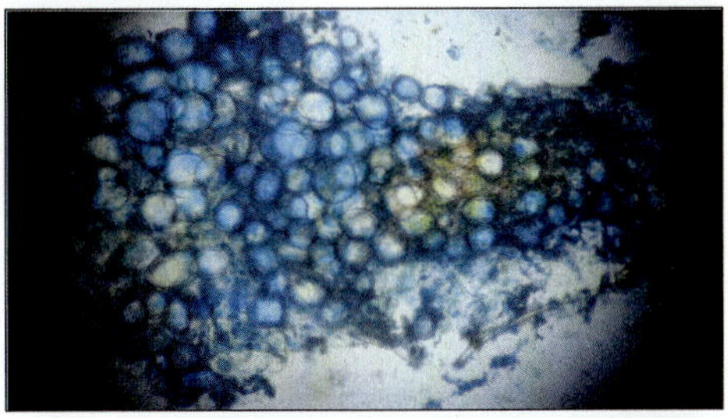

p. 120

p. 122

p. 122

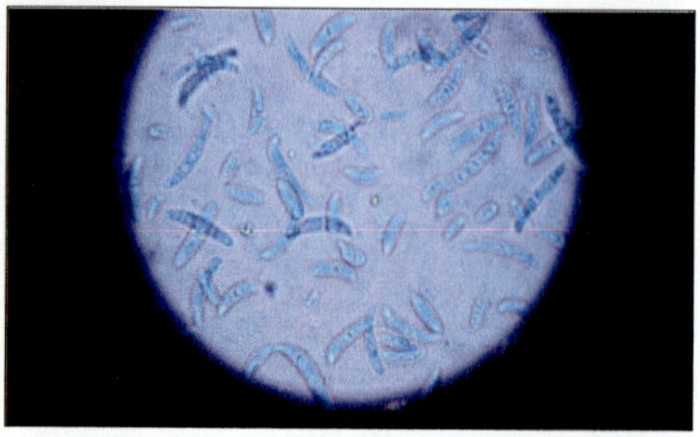

p. 123

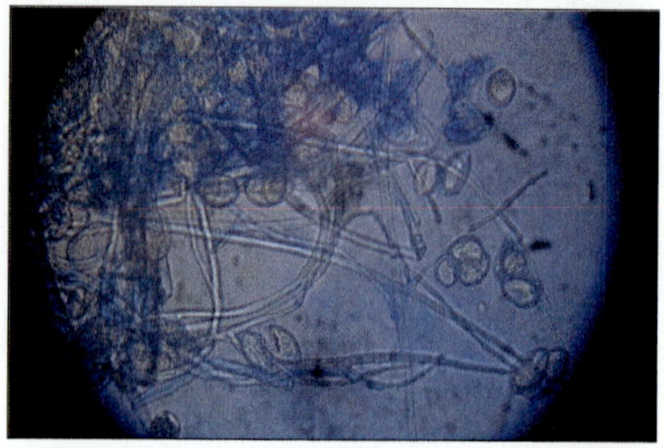

p. 125

p. 126

p. 126

p. 130

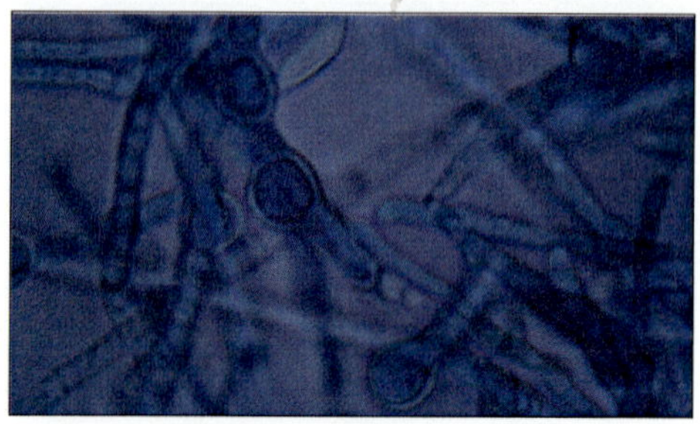

p. 130

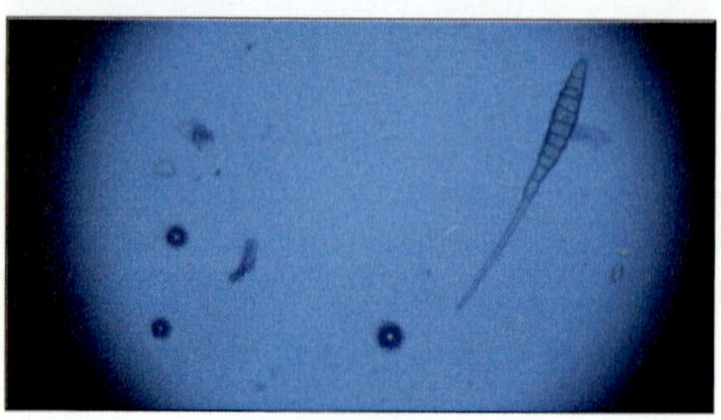

p. 137

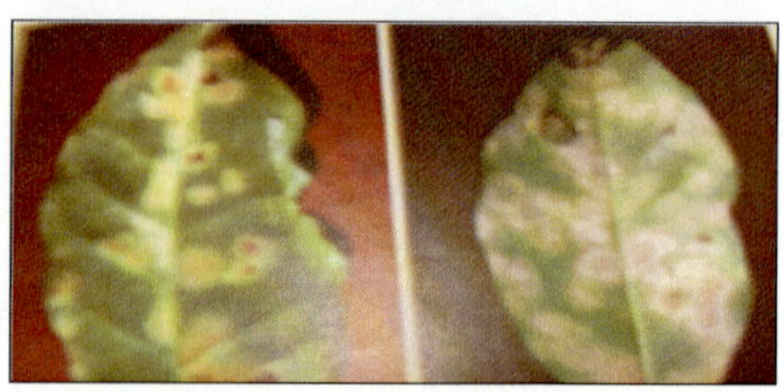

p. 146